MW01711461

Project Workbook for Systems Analysis and Design and the Transition to Objects

Sandra Donaldson Dewitz
San Jose State University

The McGraw-Hill Companies, Inc.
New York St. Louis San Francisco Auckland Bogotá
Caracas Lisbon London Madrid Mexico City Milan Montreal
New Delhi San Juan Singapore Sydney Tokyo Toronto

McGraw-Hill

A Division of The ***McGraw-Hill*** *Companies*

Project Workbook for Systems Analysis and Design and the Transition to Objects
Sandra Donaldson Dewitz

Copyright © 1996 by McGraw-Hill, Inc. All rights reserved. Printed in the United States of America. Except as permitted under the United States Copyright Act of 1976, no part of this publication may be reproduced or distributed in any form or by any means, or stored in a database or retrieval system, without the prior written permission of the publisher.

2 3 4 5 6 7 8 9 0 HAM HAM 9 0 9 8 7

ISBN 0-07-016768-0

Sponsoring Editor: Rhonda Sands
Editorial Assistant: Kyle Thomes
Production Supervisor: Natalie Durbin
Printer/Binder: HAMCO/NETPUB CORPORATION

Cover image courtesy of Quilt San Diego. Photographer: Carina Woolrich.
Quilt artist: Carol Soderlund. The quilt is titled *Covenant* © 1989.

Information has been obtained by McGraw-Hill from sources believed to be reliable. However, because of the possibility of human or mechanical error by our sources, McGraw-Hill, or others, McGraw-Hill does not guarantee the accuracy, adequacy, or completeness of any information and is not responsible for any errors or omissions or the results obtained from use of such information.

TABLE OF CONTENTS

GUIDELINES FOR USING THE CASES AND COMPLETING THE EXERCISES

This project case workbook was developed to provide systems analysis and design (SA&D) students with "hands-on" experience analyzing and designing a computer-based information system. It contains four project cases and twenty-one project exercises, each of which can be used with each case. Most exercises can be completed as individual or group assignments. They can also be completed on their own or as part of a semester-long team project. In the latter case, team members will compile their exercises as the foundation of one or more formal project reports.

By completing these exercises, you will simulate the activities of a systems analyst/designer. "Simulate" in the sense that, instead of working on a "live" project, you will work with a case problem description. Although by no means a perfect substitute for a real project, participating in this simulation will give you experience in applying the concepts and skills you learn in your systems analysis and design course. This simulation will also help you understand what a systems analyst/designer does. Besides, as an introductory SA&D student, you're not quite ready for "prime time." Believe me, as you're learning these concepts and skills, you don't <u>want</u> the added pressure of a <u>real</u> project!

Overview of the Project Cases

The four project cases describe three organizations: a non-profit clearinghouse of low-income housing information, a landscaping business, and a bookseller that specializes in rare books. All of these organizations have underutilized information technology to meet their objectives. Most perform their core processes using manual or only partly automated information system. Both the landscaping business (Natural by Design) and the bookseller (BookPlace) are inundated with paper documents, paper and notecard files, and out-of-date, inaccurate reports. The third organization -- the Homeless Coalition -- has an automated information system running on an off-site mainframe, but the users need an on-site, easy-to-use system to better meet their organizational objectives.

All of the cases assume that a team of consultants has been retained to complete the development project. YOU are one of those consultants! If you are working in teams to complete a term project, you and your team members are the consulting team. Your task is to perform the analysis and design activities of the systems development project undertaken in the case.

Most of the cases attempt to reveal the human side of information systems; that is, the personalities of users can affect how they perceive an information system or a systems development project. Some users are gung-ho, plowing ahead and offering suggestions to the consulting team. Others are reticent, either because they are shy or self-conscious about their lack of IS knowledge or -- worst case scenario -- they are opposed to the development effort. You need to be sensitive to these users as individuals, just as you would in a real-world project. Part of your job is to understand each user's perspective and to determine how best to involve each user in the project in order to maximize user commitment.

All of the cases were written with the project exercises in mind. That is, an attempt was made to ensure that each case provides most of the information you need to complete each exercise. "Most" in the sense that some exercises require you to gather data from other sources; for example, the cost-benefit analysis exercise involves contacting vendors or studying catalogs to estimate the cost of various hardware and software. If, however, you find that some case information is incomplete or inconsistent -- Welcome to the real world of SA&D! Your instructor will suggest how to deal with this problem. I tell my students that they can make whatever assumptions seem reasonable, as long as they (1) clearly indicate what their assumptions are and (2) explain how they would verify their assumptions (e.g., by consulting users).

Overview of the Project Exercises

The project exercises cover a wide range of analysis and design activities. Your instructor may assign all or only selected exercises, depending on course time constraints. Although the exercises are sequenced to align with the sequence of chapters in *Systems Analysis and Design and the Transition to Objects*, you may be asked to complete them in a different sequence. Generally speaking, Exercises 1 through 9 are designed to simulate the activities involved in completing the preliminary investigation and analysis, yielding a problem definition report. Exercises 10 through 21 are designed to simulate the activities of iterative analysis, design, and preliminary construction, yielding a design specification report.

All exercises are to be completed by studying your assigned case; in other words, the case narrative provides a context in which to perform the exercise tasks. Each exercise is designed to provide a different perspective on the case and to give you experience applying the SA&D concepts and skills you will learn in your SA&D course. If you are using *Systems Analysis and Design and the Transition to Objects* as your course textbook, you will find that, typically, each textbook chapter is supported by at least one project exercise.

Project case exercises can be completed on the forms provided at the end of the text. For many exercises, you will need to attach additional pages (e.g., providing a DFD or worksheet) to the exercise form before submitting it for instructor evaluation. If you choose, you can complete the exercises using word-processing software by using the exercise forms file provided on the project disk (see "Project Disk" below).

The Project Disk

The project disk provides several files to help you as you complete your analysis and design tasks. All of the project case exercises are included on the disk as a Microsoft Word 6.0

file (forms.doc) so that you can produce word-processed exercise solutions if you choose. In addition, a cost-benefit analysis worksheet template (template.xls in Microsoft Excel 5.0 format) is included to help you complete the related exercise. The project disk also contains sample source documents, data sets, and worksheets from the project cases. For example, Natural by Design (NBD) has created design bid and production plan worksheets using Microsoft Excel; these and NBD's inventory and supplier lists are included in the NBD-IMS directory. Similarly, the General Books division of BookPlace uses Excel worksheets to track book inventory and sales; these and excerpts of their client and source data are provided in the BP directory. In addition, Excel files of sources, clients, inventory, and the like can be imported into a database if you are assigned to implement some aspects of your system.

Approaching Your Project

Your attitude toward this simulation and the methods you bring to bear on it affect your chances of maximizing what you take away from this learning opportunity. The following paragraphs suggest how to approach this simulation, i.e., your analysis and design project.

Read Your Assigned Case Many Times. Begin by reading the assigned case as you would a short story or business article in *Fortune*. Develop a basic familiarity with the organization, the nature of its activities, and its personnel. During this first reading, avoid jumping to conclusions; at all costs, avoid deciding how to solve the organization's problems! Only after careful analysis of the organization and its objectives, the current information system and its problems, and the functional requirements of a new system will you be prepared to design a solution.

To complete the project case exercises, you must read the assigned case many times. Each exercise requires you to adopt a somewhat different perspective on the case material. One exercise may require you to read the case to identify problems with the current system; another may require you to analyze the objectives, constraints, and scope of the development project. Yet another may require you to define the system's data. Each of these analysis and design activities requires a different approach to the case. I often tell my students that, by the end of the term, they should know each case as well as they know the organization in which they work. The people, information system, business problems and opportunities should be etched on their minds by the time they hand in their second project report!

Balance Different Perspectives. Gain an overview of the exercises your instructor will assign. If your instructor has provided a course schedule, you probably know which exercises will be assigned and the sequence in which they will be completed. Scan these exercises to gain a sense of the kinds of analysis and design activities you will be asked to perform. Knowing the nature of the exercises you will complete should help you avoid trying to "do too much" in any one exercise. This knowledge will also help you prepare to wear the "many hats" of a systems analyst/designer. Some exercises require you to adopt an organizational perspective; others, a systems perspective; and still others, an information technology perspective. Don't assume that any one of these perspectives is more important than the others. The ability to view a case from all three perspectives is the defining trait of a skilled systems analyst/designer.

Sometimes a case provides different, even contradictory perspectives on a system component or organizational objective. Just as no two people in the real world hold identical views of a problem or another person (let alone of an information system!), the users in each case may describe a process differently or see opportunities where others see problems. Your job is to balance these perspectives. That is, you need to weigh each viewpoint, identify inconsistencies, and determine how best to align these different views.

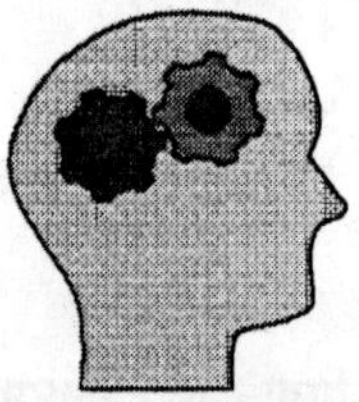

Use Your Judgment. As noted earlier, sometimes a case provides incomplete or inconsistent information. In these situations, you need to use your common sense, business knowledge, and good judgment to fill in incomplete or to resolve inconsistent information. In the real-world, systems analysts often encounter the same problems. Sometimes they are unable to get the information they need in a timely manner; sometimes interactions with users provide contradictory views. In some situations, all they can do is make an assumption, document the assumption, and ask users to verify it. You should do the same. In other situations, you may need to locate additional information, talk with other people, or apply your knowledge of similar organizations, processes, or information systems. Keep in mind that your task is to analyze and design an information system. You should bring all the resources available to this task, not just the resources provided in the case narrative.

Integrate the Knowledge Gleaned from Each Exercise. Don't treat each exercise as a stand-alone entity. Understand that exercises often build on each other. For example, the analysis of systems development goals you conducted in Exercise 1 provides insights into the system objectives you are asked to identify in Exercise 8. Similarly, the object classes and relations you define in Exercises 10 and 11 should appear as data stores in the physical DFDs you create in Exercise 12. In addition, you may be required to compile your exercise solutions to produce one or more project reports. Thus, you need to develop an integrated view of the organization, its current information system, and the proposed information system to produce a coherent, consistent analysis and/or design report.

Be Creative. Although it's probably true that solving a problem requires 98 percent perspiration and only 2 percent inspiration, that modicum of inspiration can make a mountain of difference in the quality of the solution devised. Most problems have more than one correct solution; the same is true of the project case exercises. Often times, what differentiates one correct solution from another one is the spark of creative energy the problem solver brings to the problem-solving task. Try thinking "outside the box" as you study your case and complete the project exercises.

Have Fun. You can view this simulation as an opportunity to learn what it means to be a systems analyst and to test your suitability for this profession. If you discover that you dread each exercise or team meeting, then systems analysis and design may not be a good career choice for you. If you find yourself looking ahead to later exercises, eager to undertake the new challenge each poses, then you will probably enjoy a real systems development project even more. Work consumes the majority of most adults' time and energy. If you don't enjoy your work, you probably won't enjoy your life. If you enjoy your work, you are more likely to imbue it with energy and creativity, thus enhancing your career success. This guideline is especially important if you are working in a team to complete the project. Celebrate your efforts! Have a pizza party to reward yourselves for a job well done. Be happy!

USING VISIBLE ANALYST® WORKBENCH TO CREATE DIAGRAMS

If you purchased the project case workbook shrink-wrapped with a CASE tool, you now own the student version of Visible Analyst Workbench (VAW) for Windows -- a popular, powerful, and easy-to-use tool that will ease your task of creating analysis and design models. VAW was chosen to accompany this text not only because it is easy to learn and to use but also because it supports all the modeling techniques you will apply as you complete the project case exercises.

A CASE (computer-aided systems engineering) tool is much more than a drawing tool. Although a CASE tool facilitates creation -- and modification -- of various analysis and design models, its real strength is the *intelligence* within the tool. That is, a CASE tool provides automated procedures to enforce the rules of a methodology. For example, it can verify that a child DFD is correctly balanced with its parent process; it can also analyze the complexity of a program structure chart and create system documentation.

Your copy of VAW includes a tutorial manual on using this CASE tool. This manual and the on-line help functions are your best guide to learning and using VAW. Below I discuss a few pointers to get you started using VAW and explain how to use VAW to create diagrams not specifically supported by the tool (i.e., not listed as a selectable diagram in the New Diagram dialog box).

® Visible Systems is a registered trademark of Visible Systems Corporation, 300 Bear Hill Road, Waltham, MA 02154. Telephone: 617/890-2273; Internet: http://www.visible.com.

Installing VAW

To install VAW, your computer system must be running Windows 3.1, preferably on a 486 or faster processor. A minimum of 4 megabytes of RAM is required to obtain satisfactory performance. The VAW files require approximately 4 megabytes of hard disk space, not including your project files. To install the VAW files, follow these steps:

1. Place the first VAW installation diskette into your floppy drive.
2. Select "Run" from the File menu in the Program Manager of Windows.
3. In the Run dialog box, type A:\install (or B:\install if VAW diskette is in drive b:) and select "OK" to start the installation process.
4. Follow the directions that appear on your screen to configure VAW for your system.

Creating a Project

Before you begin to create any of the diagrams for your case, ***you must define the project*** to which these diagrams belong. Defining a project creates a repository in which all diagrams are stored and related to each other and in which the rules of a methodology (e.g., Yourdon/DeMarco) are applied to the diagrams you create. When a methodology is selected, VAW automatically

- Numbers DFD processes (and data stores, if the Gane & Sarson methodology is selected); however, you can change these numbers using the "Change Item" function on the Diagram menu.
- Helps you balance child DFDs by dragging down all input and output data flows from the parent process.
- Allows you to analyze your diagrams to detect errors (see "Analyze" on the Diagram menu). For example, VAW will verify that every DFD process has at least one input and one output and that a child DFD is balanced with its parent process -- very useful features, especially for beginning DFD modelers!

Because you are using the student version of VAW, you may work with only one active project at a time. To create a project, follow these easy steps:

1. Select "New Project" from the File menu. The New Project dialog box will open, displaying text entry and other fields so that you can define the
2. Enter a name for your project (for example, 'BP' for the BookPlace case or 'IMS' of "An **I**nventory **M**anagement **S**ystem for Natural by Design"). *Project names may not exceed four characters.*
3. Enter a brief description of your project.
4. Define the path. If you are storing project files on a floppy disk, your path should be A:\ or B:\.
5. Set the repository rules governing your project.
    - Select Gane & Sarson if you plan to create DFDs using their methodology. This is the recommended methodology, assuming that you are using *Systems Analysis and Design and the Transition to Objects* as your textbook.

- Select Yourdon/DeMarco if you plan to create DFDs using their methodology.
- Select 'Disabled' if you don't want the rules of *any* methodology to be used to verify your diagrams. ***I recommend that you DO NOT select 'Disabled'; doing so will disable the verification and repository functions of VAW!***

6. You can ignore the" ERD Notation" section if you are using *Systems Analysis and Design and the Transition to Objects* as your textbook because you will not be using this modeling technique.
7. Select "OK" to save your project definition and to close the dialog box.
8. To open an existing project, select "Select Project..." from the File menu or select 🗁 from the Control bar.

Creating a Diagram

Once you have created a project, you can begin to create the diagrams needed for the project. VAW supports creation of all the modeling techniques required for your project case exercises: data flow diagrams, system flowcharts, workflow diagrams, organization charts, object relationship models, and program structure charts. Because you are using the student version of VAW, the number of each diagram you can create is limited to ten; that is, you can create 10 DFDs, 10 structure charts, and so on.

General directions for creating a diagram are provided below. More specific information is provided for those diagrams not explicitly available in the New Diagram dialog box.

1. Select "Open Project" from the File menu or select the Open Project icon (🗁) from the Control bar to open a project. The Select Project dialog box will open.
2. Select the project for which you plan to create a diagram; select "Select." The Select Project dialog box will close.
3. Select "New Diagram" from the File menu or select the New Diagram icon from the Control bar. The New Diagram dialog box will open, providing options for you to select the type of diagram you wish to create.
4. Select the desired diagram type from the Diagram Type combo box. You can safely ignore the other selection criteria for most diagrams.
5. Select "OK" to close the dialog box.
6. To display the default symbols of your chosen diagram, select the Add Symbol icon from the Control bar or select "Symbol" from the Diagram menu. To open the toolbox to see all symbols available, select "Toolbox" from the Options menu.
7. To add a symbol to your diagram, select the symbol from the toolbar or toolbox; the cursor will change from an arrow to a ✸ (similar to a ship's wheel).
8. Position the ✸ where you want the symbol to appear; click the left mouse button. The selected symbol will be placed in your diagram, and the Label Object... dialog box will appear.
9. Enter a label for your symbol; select "OK" to close the dialog box.

Creating a System Flowchart. Perform the steps described above for creating a new diagram, selecting "Unstructured" as the diagram type. Open the toolbox to display all the symbols available. Use the up and down arrows of the toolbox to examine the symbols available. Note that the most commonly used system flowchart symbols are provided here. Use the relevant symbols in the toolbox to create process, DASD, SASD, document, and other symbols in your flowchart. As you add these symbols to your diagram, the Label Object... dialog box described in the general directions above will open so that you can label these symbols. Use the line symbol in the toolbox to create directional lines between symbols.

<u>Creating a Workflow Diagram</u>. Perform the steps described above for creating a new diagram, selecting "Unstructured" as the diagram type. Open the toolbox to display all the symbols available. Use the plain circle and plain square symbols in the toolbox to create internal and external entities, respectively. As you add these symbols to your diagram, the Label Object... dialog box described in the general directions will open so that you can label these entities. Use the line symbol in the toolbox to create workflows. Select and double-click on a workflow line to label your workflow.

<u>Creating an Organization Chart.</u> Perform the steps described above for creating a new diagram, selecting "Unstructured" as the diagram type. Open the toolbox to display all the symbols available. Use the plain square symbol in the toolbox to create organizational entities. As you add these symbols to your diagram, the Label Object... dialog box described in the general directions will open so that you can label these entities. Use the line symbol in the toolbox to create lines of control between entities.

<u>Creating an Object Relationship Model</u>. Perform the steps described above for creating a new diagram, selecting "Class" as the diagram type. Follow the instructions provided in your tutorial manual to create object classes and relationships. Note that the notation for cardinality is different from that used in *Systems Analysis and Design and the Transition to Objects*. Instead of using crow's feet and so on to indicate cardinality, VAW uses the symbols shown below mapped to the textbook notation:

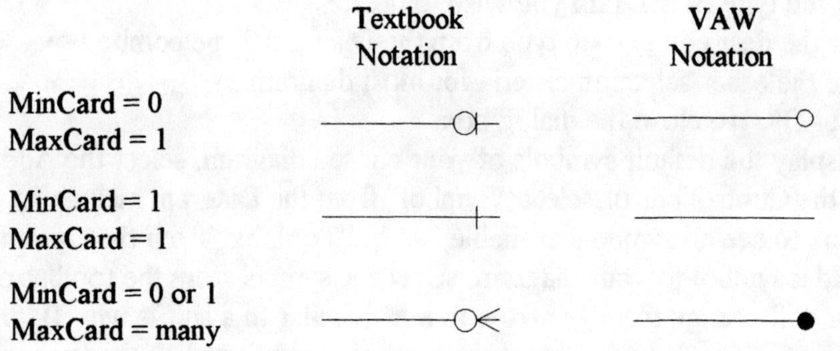

<u>To add class attributes</u>, select a class in your diagram, and then select "Define..." from the Repository menu. The Define Item dialog box will open. Place the cursor in the Attributes box, and select the QComp button at the bottom of the dialog box. A dialog box labeled "Attributes for <class name>" will appear. Define each attribute (select the Help button in the Attributes dialog box to see definitions of attribute properties, e.g., Type, Limit, and so on). When you have

entered all the attributes for a particular class, select the Next button to define attributes for another class. When you are finished entering attributes, select the Exit button to return to your diagram. The attributes you defined now appear in the appropriate classes.

To add class methods, select a class in your diagram, and then select "Define..." from the Repository menu. The Define Item dialog box will open. Select the page forward button (**»**) until the page of the class definition for methods appears. Place the cursor in the Methods box, and select the Qcomp button at the bottom of the dialog box. A dialog box labeled "Method definition for <class name>" will appear. Define each method (select the Help button in the Methods Definition dialog box to see definitions of method properties). When you have entered all the methods for a particular class, select the Next button to define methods for another class. When you are finished entering methods, select the Exit button to return to your diagram. The methods you defined now appear in the appropriate classes.

A HOUSING INVENTORY SYSTEM FOR THE HOMELESS COALITION

I. Background

The Homeless Coalition is a nonprofit organization that advocates the development and improvement of low-income housing as a means to reduce homelessness. The Homeless Coalition operates in a largely metropolitan county of approximately 1.5 million people in Northern California.

The Coalition serves as an information clearinghouse, providing three basic services. First, it works with politicians, lobbyists, and advocacy groups to foster legislation that encourages the development of low-income housing through tax incentives, developmental zoning preferences, and other legislative inducements. To accomplish this, the Coalition provides information about low-income housing to state, county, and city governments. Second, the Homeless Coalition strives to raise community consciousness about the need for low-income housing through speeches, seminars, displays at conventions, and other public relations activities. Finally, the Coalition provides information about the availability of low-income housing to other agencies that work with the low-income and homeless populations.

A. Staff and Facilities

The Coalition employs a director, a demographer, a statistician, three full-time administrative assistants, and a varying number of unpaid part-time volunteers. Its office is located in the basement of a building owned by a developer who supports the coalition's work. Currently, the Homeless Coalition has limited information technology resources: two dumb terminals that provide time-sharing access to the county's mainframe. The statistician is the most computer literate of the Coalition's employees; she is able to run SPSS statistical models on the IBM 3090.

The other employees are intimidated by the log-on procedures, command language, and character-based interface of the VM operating system. Thus, whenever they need reports run on the housing and demographic data stored in DB2 tables on the mainframe, they put in a request to the county's Information Services staff -- and wait for a printout to be delivered. This arrangement is less than ideal because, typically, the turnaround time is 24 hours for a printout of an existing report and weeks or months for a new report. New reports take so long because the already overburdened IS staff must write a program to generate the requested report.

The Coalition's budget is supported by funds from several sources, including HUD, the county, a community services fund, and several businesses. The director of the agency, Brenda Campsey, must reapply for funds from her sponsors each year. Brenda would like to expand Coalition activities, but is unable to do so due to limited funds. Brenda believes that a number of the businesses in the community would be willing to provide additional funds if she could demonstrate its effectiveness as an information resource. Unfortunately, she has been unable to find time to present these ideas to the businesses or even to develop such a presentation.

B. Data

As noted, all of the data the Homeless Coalition needs is stored on the county's mainframe system. Data stored includes the location, availability, and condition of low-income housing in 50 census tracts[1] within the county. These census tracts were chosen because data indicate that, of the approximately 300 census tracts in the county, these fifty have the highest percentage of their populations living below the poverty level. Within these 50 tracts are approximately 200 multiple-family and 350 single-family low-income housing facilities. On average, each multiple-family facility contains 30 units.

The Coalition maintains data about each census tract, including tract number (designated by the U.S. Census Bureau), geographical boundaries, median household income, income per capita, median household income for renters, per capita income for persons in group quarters (e.g., a rooming house), median persons per rental household, median gross rent, and median gross rent as a percentage of rental family income. The coalition also maintains the name, address, and phone number of elected officials and principal property developers in each tract. Each single-family facility is described by its address, square feet, number of bedrooms and baths, outdoor living space (e.g., fenced backyard, patio), renovations and repairs, and accessibility by the handicapped. For each multiple-family facility, the building name, address, number of units, manager name and phone number, accessibility by the handicapped, and ratings of the building's facilities are maintained. Each unit (e.g., apartment) in the multi-family buildings is described by its unit number, square feet, number of bedrooms and baths, availability of cooking facilities,

---

[1] A census tract is a geographical area designated by the U.S. Census Bureau for the purpose of data collection. On average, each census tract has a population of 4,000.

occupancy status, and any special remarks. Data about the single- or multi-family building's owner are also tracked, including owner name, address, and phone number.

II. The Housing Inventory Application

A. Current System

The county's IS staff have supported the Homeless Coalition by developing and maintaining its Housing Inventory Application (HIA), which regularly produces three reports. Figure HC-1 shows the Census Tract Report, which lists census tracts, the buildings located in those tracts, and the units located in each building. The number of units and the average monthly payment are shown for each building, for each census tract, and for the entire region. (Note: All figures show only a sample of the data.)

CENSUS TRACT REPORT

Census Tract Number		10058	Per Capita Income	$13,567.00
Building Name Elm St.			Building Owner C. Lyons	
Unit#	Bedrooms	Size	Rent	
101	0	245	175.00	
102	1	380	305.00	
104	2	510	425.00	
206	1	398	320.00	
....	Total number of units		25	
	Average rent per unit		$361.25	
Building Name Third Bldg			Building Owner C. Lyons	
Unit#	Bedrooms	Size	Rent	
200	1	398	276.00	
202	1	387	288.00	
....	Total number of units		30	
	Average rent per unit		$352.00	
.....	Number of units for this census tract			90
	Average rent for units in this area			$357.83
Census Tract Number		10078	Per Capita Income	$14,879.00
Building Name 14th Ave			Building Owner P. Peacock	
Unit#	Bedrooms	Size	Rent	
10	0	267	158.00	
20	0	287	175.00	
30	0	321	271.00	
....	Total number of units		30	
	Average rent per unit		$278.00	
.....	Number of units for this census tract			159
	Average rent for units in this area			$389.47
			Total number of units	6376
			Average rent per unit	$366.00

Figure HC-1: Census Tract Report

The second report is the Unit Type Report (Figure HC-2), which lists units by the number of bedrooms. (An efficiency unit is shown as having zero bedrooms.) Within each category, the report lists buildings that contain units of that type as well as data about the individual unit, including whether or not the unit is occupied ("O" signifies occupied; "V" signifies vacant).

UNIT TYPE REPORT

Number of bedrooms: 0

Building Name	14th Ave		Building Owner	P. Peacock
Unit#	Size	Rent	Occupied Code	
10	267	158.00	O	
20	287	175.00	O	
30	321	271.00	V	
.....				
Total number of units of this type in this building			6	
Total number of available units in this building			1	
Average rent per unit of this type			$208.00	

Building Name	Elm St.		Building Owner	C. Lyons
Unit#	Size	Rent	Occupied Code	
101	245	175.00	O	
.....				
Total number of units of this type in this building			5	
Total number of available units in this building			0	
Average rent per unit of this type			$175.00	

Number of units with 0 bedrooms	378
Number of units available	13
Average rent for these units	$217.25

Number of bedrooms: 1

Building Name	Elm St.		Building Owner	C. Lyons
Unit#	Size	Rent	Occupied Code	
102	380	305.00	O	
206	398	320.00	O	
.....				
Total number of units of this type in this building =			15	
Total number of available units in this building =			0	
Average rent per unit of this type =			$292.50	

Building Name	Third Bldg		Building Owner	C. Lyons
Unit#	Size	Rent	Occupied Code	
200	398	276.00	O	
202	387	288.00	O	
.....				
Total number of units of this type in this building			15	
Total number of available units in this building			1	
Average rent per unit of this type			$282.00	

Number of units with 1 bedrooms	1871
Number of units available	26
Average rent for these units	$298.50

Figure HC-2: Unit Type Report

Number of bedrooms: 2

Building Name		Elm St.	Building Owner	C. Lyons
Unit#	Size	Rent	Occupied Code	
104	510	425.00	V	
.....				

Total number of units of this type in this building 5
Total number of available units in this building 1
Average rent per unit of this type $425.00

.....

Number of units with 2 bedrooms 2793
Number of units available 21
Average rent for these units $405.00

.....[2]

Total number of units 6376
Total number of available units 83
Average rent per unit $366.00

Figure HC-2: Unit Type Report (continued)

The third major report, the Owner Report, is illustrated in Figure HC-3. This report is sorted by owner name and lists the buildings and units provided by each owner. This report also shows the quality ratings for each building and a composite rating on each criterion for each owner.

B. The Housing Inventory System Project

Recently, the Homeless Coalition applied for a grant to develop an on-site Housing Inventory System (AIS). The HIS would provide a more user-friendly method to access and manipulate data so that Coalition workers could more easily fill requests for information. The plan is that the staff themselves should be able to produce the reports that their clients -- politicians, lobbyists, advocacy groups, and agencies supporting the low-income population -- need to foster the growth and to exploit the availability of low-income housing in the county. This capability is critical to the Coalition's mission as an information clearinghouse. In addition, the HIS would allow them to import data into spreadsheets for analysis and to create their own promotional literature (e.g., mass mailings to businesses requesting financial support, brochures describing services, etc.).

[2] Units with three or more bedrooms are included in the Unit Type Report but are not shown here.

OWNER REPORT

Owner Name C. Lyons — Owner Addr Trump Tower, Suite 2000, New York, NY 10022

Building Name Elm Street

Unit#	Bedrooms	Size	Rent
101	0	245	175.00
102	1	380	305.00
104	2	510	425.00
206	1	398	320.00
.....			

Total number of units 25
Average rent per unit $361.25

Quality Ratings (Scale: 1-poor to 5-excellent):

Building upkeep	4.0
Tenant service	4.0
Heating/Cooling system	4.0
Pest control	4.0
Average rating:	4.0

Building Name Third Building

Unit#	Bedrooms	Size	Rent
200	1	398	276.00
202	1	387	288.00
....			

Total number of units 30
Average rent per unit $252.00

Quality Ratings (Scale: 1-poor to 5-excellent):

Building upkeep	5.0
Tenant service	4.0
Heating/Cooling system	4.0
Pest control	4.0
Average rating:	4.25

.....

Owner Summary: C. Lyons
Total number of units 326 — Average rent $315.00 — Composite rating 4.125

Owner Name P. Peacock — Owner Addr 13530 First Street, Suite 200, New York, NY 10021

Building Name 14th Ave

Unit#	Bedrooms	Size	Rent
10	0	267	158.00
20	0	287	175.00
30	0	321	271.00
.....			

Total number of units 30
Average rent per unit $278.00

Quality Ratings (Scale: 1-poor to 5-excellent):

Building upkeep	5.0
Tenant service	4.0
Heating/Cooling system	4.0
Pest control	5.0
Average rating:	4.5

.....

Owner Summary: P. Peacock
Total number of units 143 — Average rent $318.00 — Composite rating 4.5

Summary for All Owners

Total number of units 6376
Average rent $366.00
Quality Ratings:
(Scale: 1-poor to 5-excellent)

Building upkeep	3.7
Tenant service	3.5
Heating/Cooling system	4.0
Pest control	4.4
Average rating:	3.9

Figure HC-3: Owner Report

Last month, HAL, Inc., a major computer company with headquarters in the county, approved the coalition's grant and agreed to fund the development of the HIS. However, HAL limited the scope of the development effort to creating the data files and generating the three major reports outlined in the grant (see Figures HC-1 through HC-3). In addition, HAL will provide six personal computers, a server, three laser printers, and whatever other hardware and software the Coalition needs to set up its own LAN system. The current mainframe HIA data will be redeployed to a LAN running Windows for Workgroups and Microsoft Office.

III. JAD Workshop with Coalition Workers

As part of its grant, HAL also provided a small team of consultants to analyze the Coalition's needs and to design and implement the HIS. The team held a two-day JAD workshop with coalition employees and one member of the county's IS staff to determine requirements and to outline the design of the new system. Portions of this workshop are described here.

A. Organizational Structure and Workflows

"We have a general overview of your role as an information clearinghouse, but we need to gain a better understanding of how each of you interacts with other Coalition employees and volunteers and how the Coalition interacts with outside agencies. Brenda, could you describe the Coalition's organizational structure for us?"

"Sure," Brenda responded. "Being such a small organization, we have a pretty informal structure. As director, I oversee all coalition activities, including hiring employees, maintaining our financial records, and assigning duties to employees. I also do most of the public relations work, interacting with our sponsors and clients on a face-to-face basis. My staff includes a statistician, Bob Franklin; a demographer, Ginny Alvarado; and three assistants -- Bonnie Devine, Theresa Yoon, and Steven Lam. In addition, our staff is supplemented by a varying number of volunteers -- currently about six -- who perform various duties, including filing, conducting telephone phone interviews and fund solicitations, processing mail, and so on. Theresa recruits and supervises these volunteers; she also is responsible for gathering and maintaining data about and planning the campaigns to solicit contributions from our sponsors: the agencies, businesses, and individuals who fund our operations."

A consultant interrupted, "Would an appropriate title for Theresa's position be fund-raising manager?"

"I suppose," Brenda replied. "But, as I noted, she has other duties as well. Perhaps 'Community Support' would be a good tag for what she does. That title includes all aspects of recruiting volunteers and soliciting financial support from the community -- basically interfacing with our sponsors. Steven is primarily responsible for collecting and maintaining data about low-income facility owners and the facilities they own. He is our interface with owners and facility managers -- also with the city and county building inspectors who provide us information about the quality of these facilities. Bob, do you have a copy of the form you use to collect these data?"

"Yes," Bob answered. "Here's the Unit Data Collection form we ask the building manager or owner to complete when we first learn about the low-income housing facilities (see Figure HC-4). Then, on the first of each month, the manager or owner submits a report indicating occupancy status and rent changes, if any (see Figure HC-5). In addition, twice each year, the county or city building inspector performs an unannounced on-site evaluation of each low-income facility and sends us a copy of the evaluation" (see Figure HC-6).

MULTI-FAMILY LOW-INCOME UNIT DATA COLLECTION FORM

Building Name ____________________ Building Address ____________________
Owner Name ____________________ Owner Phone ____________________
Owner Address ____________________ Owner Occupied? _____ Yes _____ No
Manager Name ____________________ Manager Phone ____________________

Please provide the following information about each unit in your low-income housing facility.

Unit #	Number of Bedrooms	Rent	Square Feet	Cooking Facilities (Y/N)	Number of Baths	Specific Remarks	Occupied? (Y/N)

Figure HC-4: Unit Data Collection Form

MULTI-FAMILY UNIT DATA CHANGE REPORT
for March 1996

Building Name Elm Street Apartments Building Address 132 Elm Street, San Jose
Owner Name C. Lyons Owner Phone 212/333-5423
Owner Address Trump Tower #2000, NY 10022 Owner Occupied? _____ Yes ✓ No
Manager Name Don Calhoun Manager Phone 408/297-5511

Please indicate any changes in the owner, manager, and unit information for your housing facility.

Unit #	Old Rent	New Rent	Changes in Conditions	Last Month Occupancy	This Month Occupancy
101	$175	$185		V	O
216	$255	$270		O	O

Figure HC-5: Unit Data Change Report

City of San Jose
Building Inspection Report
for Inspections Completed March 1996

Building Name	Address	Upkeep	Service	H/C System	Pest Control
Elm Street Housing	123 Elm St.	4.0	4.0	4.0	4.0
Third Bldg Apts.	1045 Third Ave	5.0	4.0	4.0	4.0
14th Ave Vista	500 14th Ave	5.0	4.0	4.0	5.0
.....					

Figure HC-6: Building Inspection Report

"Thanks, Bob," Brenda continued. "I guess the only assistant I haven't talked about is Bonnie, who is unable to be with us today. Bonnie focuses on our clients -- the city, county, and state agencies that use the data we collect. She fields questions from these clients, determines the kinds of reports we need to generate to meet their needs, and, consequently, the data we need to collect to generate these reports. As we receive printed reports from the county IS staff, Bonnie determines which agencies will be interested in each report, makes photocopies, and sends the copies to the interested agencies."

"Okay, that's pretty clear cut. But some of the tasks you've described your assistants performing seem like pretty sophisticated work? Are you sure that 'administrative assistant' is an appropriate job title?" a consultant asked.

"We're not too hung up on titles around here," Theresa said. "We do the work that needs to be done to help provide decent low-rent housing. Besides, we assistants don't do any of the heavy data manipulation or number crunching. We just collect the data and transmit reports to outside agencies; Ginny, Bob, or the county IS people do the 'sophisticated' work -- as you call it. They compile the data to produce the reports."

"Perhaps Ginny could describe who provides her and Bob with what information. That may help you understand our internal operations," Brenda noted. "Ginny, why don't you explain what you do."

"My major task," Ginny began, "is to compile Census Bureau data about each census tract. In addition, I occasionally lead a small group of volunteers in canvassing selected areas, collecting information about the number of homeless people, how long they've been homeless, the reasons for their homelessness, et cetera. I guess more to the point is the work I do to demonstrate the need for low-cost housing. This county is among the top three in the U.S. in terms of its housing costs. Its median family income is also quite high -- just over $50,000, and the per capita income is more than $20,000 in some census tracts.

"Unfortunately, the median family and per capita income of many census tracts in our county is skewed in the sense that the top half of the population earns a very nice living in high-tech jobs. For this top half, their income is significantly higher than the U.S. average. But we also have the bottom half whose income is significantly lower than the U.S. average. For example, census data indicate that the median family income for rental households is about $36,000; furthermore, the per capita income of persons living in group quarters is about $7600. So, if we looked only at the bottom third of the population, the income in many census tracts would be much lower than the figure indicated. Can you imagine a family of four with an annual gross income of $25,000 being about to afford the typical apartment in this county? At $755 per month in most of our census tracts, more than half of that family's take-home income would go to rent."

One of the consultants summarized Ginny's comments. "Okay, then, in terms of workflows, you gather data yourself, from the Census Bureau, or from other sources about the per capita income, the cost of housing, the demand for low-income housing, and so on in each census tract. What do you do with this data?"

"I usually give it to Bob, our statistician, so that he can analyze it to spot trends, inconsistencies, et cetera. Bob couldn't be here today, but I often work jointly with him, so I can tell you about his work. We take the building and unit data that Steve collects and generate summary statistics such as the average rent per low-income unit in each census tract (see Figure HC-1), the number of units of a particular type -- efficiency, one-bedroom, and so on -- in each census tract, and that type's average rent in each census tract (see Figure HC-2). We also track the quality of low-income housing, as indicated in the Owner Report (see Figure HC-3). We'd like to track average occupancy rates as well to get a feel for where the demand is -- in one bedroom apartments, two-bedroom houses, whatever. So far, we've been unable to track this information very well.

"Although not shown in the sample reports, we also collect Census Bureau data showing the average cost of non-low-income units of each type in each census tract -- the $800 per month that I mentioned earlier -- and relating that average cost to what low-income families can actually afford to pay for rent. The data that I and volunteers collect about the number of homeless people or families in the county is compared to the number of low-income units available. Typically, we show that, assuming these people could afford the rent on these low-income units -- which many of them could, there are not enough units available to meet the need."

"Is that data also maintained on the county mainframe?" one of the consultants asked. Ginny shook her head. "If not, then we may be getting off track here. What you are describing is very interesting, Ginny, but we need to focus on just your current, regularly generated reports. Our charge is to set up an automated on-site information system to maintain the data required to generate your three major reports and to actually produce these reports. We'll also train you to use the new system so that you can generate any other reports you may need any time you need them."

B. The Mainframe Connection

Bill Garvey from the county's IS staff also attended the JAD workshop. Bill has been responsible for the Coalition's HIA system since it was first developed in the late 1980s.

"I'd have to say that there's nothing fundamentally wrong with the HIA as it currently exists," Bill stated. "I understand Brenda's frustration with the reporting situation, but that's a logistical problem -- and a user problem, not an IS problem. I've offered to have a line printer installed here in the Coalition office so that they would have ready access to reports. I've also offered to train anyone interested in learning SQL so that they could generate ad hoc reports. I've had no takers though."

"No one's criticizing you or the HIA, Bill," Brenda interrupted. "The issue is broader than you make it out to be. Not only do we need a more easy-to-use system; we also need a more flexible system so that we can perform the full range of activities required to operate this agency: not just reports, but word processing, accounting, client management, fund raising, statistical analysis --"

"I know, Brenda," Bill interrupted. "I'm not being defensive. I'm just trying to make clear that studying the existing HIA provides a basic model of at least the data management and report generation functions of the HIS."

"We've made some efforts to more fully utilize the HIA, Bill. Bob has trained Steve to update the building-rating and unit tables."

"That's right, Brenda, but only after I reminded you for the umpteenth time that I can't generate accurate, up-to-date reports from inaccurate, out-of-date data."

"And that's where the problem lies. No one at the county has time to maintain our data, and we have neither the time nor the expertise to do so ...."

One of the consultants interrupted Brenda. "Let's bury the old HIA, okay? Whatever your problems were, however you may feel about the HIA, it's a done deal that it will be replaced by the HIS. So let's not get bogged down in last week's news, okay? Bill, please explain what you mean about using the existing HIA as a model for the HIS."

"Basically, the Coalition is an I/O operation. They gather a variety of inputs and then sort, aggregate, summarize those inputs to produce reports. There's no real transaction processing, no real need for real-time processing. For example, currently we generate their reports as regularly scheduled batch jobs. As far as developing the HIS is concerned, the easy part of your job is developing some easy-to-use data entry screens for file updates and creating the programs to generate the three major reports. The more difficult job is designing general query or report definition screens and training Coalition users to use them to create their own queries and to generate their own reports. Some kind of on-line analytical processing (OLAP) application seems to be needed. I can give you printouts of code and reports for all of the other reports Brenda has had us generate over the years. Those may give you an idea of the kinds of queries the HIS will need to support."

"We'll keep that in mind, Bill. But, again, I must remind you that HAL is funding development of only the basic system, i.e., the three major reports, and user training. Perhaps after we've done our job, Brenda can get the county to set up the other reports you've mentioned on the new HIS."

AN INVENTORY MANAGEMENT SYSTEM FOR *NATURAL BY DESIGN*

I. Background

Natural by Design, Inc. (NBD) is a partnership that designs and develops interior and exterior landscapes for hotels, restaurants, condominiums, and private estates. In their designs, they use a mixture of commonly available plants, flowers, and shrubs; imported rare and exotic plants; pottery; and other materials. NBD employs thirty people and has gross sales of $4.1 million with a before-tax profit of $750,000. Keri Yamaguchi and Stan Fenton own the partnership equally. Keri, a designer, manages the artistic side of the business -- including supervising the design activity and job production and selecting vendors. Stan is the general manager who purchases inventory and supervises the sales and financial activities.

A. Acquiring Projects

To acquire projects, the NBD salespeople call upon potential customers and identify project opportunities. Once the salespeople identify a real prospect, they introduce one or more of NBD's five designers to the client. The designers meet with the client, determine the client's needs, and then develop a design, including formal sketches and blueprints. The designers also prepare a written Design Bid (see Figure IMS-1) that contains a general description of the project, a list of plants and materials, and a detailed listing of both material and labor costs. The design bid and sketches are reviewed with the customer, adjusted as needed, and then approved (or rejected). If approved, the project is turned over to the Production staff for implementation. Accounts Receivable then follows up for payment. Usually, customers are slow to approve a design, but, once they do approve it, they want near immediate implementation. Because NBD's procurement process is very slow, the designers have learned to work only with plants and materials that are in inventory or that will very soon be in inventory. Except for very large and specialized jobs, the designers do not have the luxury of selecting plants that are not already on order or in inventory.

Natural by Design

14398 Union Road   San Jose, CA 95190

(408) 775-3600

Design Bid

Client

Client Name:	London Sq. Mall	Contact:	Amy Benson, Mgr.
Client Address:	12638 Mall Drive	Phone:	(408) 834-5603
	Scotts Valley, CA 95066		

NBD Staff

Sales Assoc:	Bill Reinhardt	Phone:	(408) 775-3652
Designer:	Tamara Bakken	Phone:	(408) 775-3645

Project

Bid Date:	May 6, 1996	Project Site:	Main entrance
Est. Begin Date:	June 15, 1996		Mall Dr./Cinema Lane
Est. Compl. Date:	June 30, 1996	Bid Amount:	$7,651.50

Material Requirements

Plants:				
Qnty	**Desc**	**Size**	**Unit Price**	**Extended Price**
3	laccospadix australasica palm	15 gal	$749.00	2,247.00
5	caryota mitis	7 gal	$233.00	1,165.00
7	marginata	2 gal	$75.00	525.00
Pottery:				
1	granite fountain (GFN48)	48 in	$750.00	750.00
3	granite pots (GP50)	50 gal	$195.00	585.00
Materials:				
10	decorative cedar bark (CBRK5)	5 cu ft	$15.95	159.50
1	top soil (TSOIL)	yard	$20.00	20.00

Labor Requirements

Hours	**Desc**	**Unit Price**	**Extended Price**
30	Production workers	$30.00	900.00
10	Design consultant	$65.00	650.00
10	Heavy equipment operator	$65.00	650.00

Figure IMS-1: Design Bid

B. Managing Inventory

Effective inventory management is critical to NBD's success. They must have appropriate plants and other materials available when needed; however, because their inventory is expensive to maintain, they do not want more inventory than necessary. But with order lead time averaging 5 to 6 weeks on some of their more exotic plants, they need to maintain sufficient stock to carry them until the ordered items are delivered.

The Design Group keeps inventory records to facilitate the effective use of inventory. Specifically, for each plant type, they maintain data about the plant type, size, average net cost, list price, quantity in stock, quantity in stock and on bid, quantity on order, and quantity on order and on bid. The on-bid quantities are tracked so that designers know the quantity of uncommitted inventory available at a point in time. Similar information is maintained about pottery and decorative items, but only in stock and in stock/on bid quantities are maintained. For landscaping materials, only the description, unit, net and list prices, and quantity in stock are recorded. An excerpt of this inventory data is shown in Figure IMS-2.

NBD Inventory Report **June 13, 1996**

PLANT INVENTORY

Code	Desc	Size	Last Ordered	AveNet$	List$	QIS	IS/OB	QOO	OO/OB
lacco	lacco australasica	15 gal	06/11/96	450.00	749.00	3	3	3	1
arenga	arenga pinnata	15 gal	05/15/96	310.00	516.00	4	2	3	0
cham	chamaedorea	15 gal	06/09/96	300.00	499.00	0	0	2	0
cera	ceratozamia molongo	14 in	05/30/96	240.00	400.00	4	2	6	0
areca	arecastum coco	15 gal	05/15/96	275.00	458.00	11	5	0	0
cary	caryota mitis	7 gal	05/30/96	140.00	233.00	14	9	4	0
grnti5	green ti	5 gal	05/30/96	92.00	154.00	11	8	6	0
grnti7	green ti	7 gal	06/09/96	140.00	234.00	8	5	6	0
ficus14	ficus green gem	14 in	06/12/96	90.00	150.00	7	7	20	2
ficus17	ficus green gen	17 in	06/12/96	240.00	400.00	6	4	12	0
margi	marginata	2 gal	06/11/96	45.00	75.00	16	13	12	0

POTTERY INVENTORY

Code	Desc	Size	Features	AveNet$	List$	QIS	IS/OB
TCP50	t/c pot	50 gal	hand painted	53.95	110.95	6	3
GP50	granite pot	50 gal	oblate	110.00	195.00	5	3
TCF03	t/c figurine-swan		glazed	25.50	45.00	3	1
MBB30	marble bird bath	30 in	Doric base	128.50	250.00	2	1
GFN48	granite fountain	48 in	fluted basin	457.50	750.00	1	1

MATERIALS INVENTORY

Code	Desc	Unit	AveNet$	List$	QIS
CBRK5	decorative cedar bark	bag (5 cu ft)	7.50	15.95	53
CRGRN	crushed granite	yard	7.50	14.00	12
PGRV	pea gravel	yard	8.00	20.00	7
GRV1	1" gravel	yard	5.90	12.00	18
TSOIL	top soil	yard	12.50	20.00	10
PBLKG	patio block-gray	each	.56	.84	94
PBLKR	patio block-red	each	.56	.84	123

Figure IMS-2: Excerpt from NBD Inventory Report

Inventory information is maintained as an Excel worksheet by the Design Group's administrative assistant, Connie Nguyen. Each day, Connie updates the numbers, prints the worksheet, and makes a copy for each designer. The designers work against the data on these copies in deciding which plants to include in their designs.

There are a number of problems with this arrangement. First, designers work simultaneously, and sometimes they unknowingly use the same last items from inventory in their bids. They try to coordinate their activities to prevent this, but sometimes it is not possible. Second, if a design bid is rejected or changed to eliminate plants or other materials, these items should be placed back in inventory; that is, Connie should subtract them from the appropriate on-bid quantity total and add them to the in-stock or on-order quantity total. Sometimes designers forget to tell Connie about these changes. As a result, the number of items available to be used in bids is often larger than the records show.

C. Purchasing and Maintaining Stock Items

Stan Fenton is responsible for ordering stock, but because he lacks Keri's knowledge of plants, he relies on her expertise to select suppliers for all NBD's plant stock. Once a year Keri evaluates the price and quality of the plants and the quality of service provided by each plant supplier. She compiles this information into a report that lists the preferred and secondary supplier for each plant type; also listed are the total quantity of each plant type used by NBD in the previous 12 months and each supplier's order lead time, i.e., the average length of time between NBD's placing an order and receiving the corresponding shipment. (NBD does not need to maintain similar information for its other pottery and materials. These items are obtained from local suppliers who can usually deliver an order within 48 hours.) An excerpt from this report is shown in Figure IMS-3.

Description	Size	Sales/Yr	Supplier1	Supplier2
laccospadix australasica palm	15 gal	100	Palms(4)	Lunds(5)
arenga pinnata palm	15 gal	160	GPC(2)	Greens(3)
chamaedorea falcifera palm	15 gal	200	Palms(4)	Lunds(5)
ceratozamia molongo palm	14"	175	Palms(4)	Lunds(5)
arecastum coco plumosa	15 gal	265	Greens(3)	GPC(2)
caryota mitis	7 gal	350	Olsen(3)	Greens(3)
green ti	7 gal	400	TisRUs(5)	Olsen(6)
green ti	5 gal	600	TisRUs(5)	Olsen(6)
ficus green gem	14"	500	Figs(4)	Lunds(5)
ficus green gem	17"	350	Figs(4)	Lunds(5)
marginata	2 gal	500	GPC(2)	Greens(3)

Figure IMS-3: Suppliers, Lead Times, and Sales for Each Plant Type

Stan refers to this report when he is preparing a purchase order to determine what to order. He uses the annual sales quantity figure to determine the average sales per week of each plant; then he compares this figure to the current stock level, taking into consideration the time it will take to restock each plant type. For example, on June 12 he noted that the current stock levels of the 14-inch and 17-inch ficus were only 7 and 6, respectively. Given the information that, on average, NBD sells 4 of the 14-inch and 2 of the 17-inch plants each week and that the order lead time is 4 weeks, Stan rushed out an order for 20 of the 14-inch ficus and 12 of the 17-inch.

D. Tracking Inventory Transactions

Each day, Connie receives a number of forms that she must compile to update the inventory report. At the end of each day, the designers give Connie a copy of any Design Bids they have prepared (see Figure IMS-1), which list the materials they plan to use. Connie adds these item quantities to the in stock/on bid (IS/OB) or on order/on bid (OO/OB) totals, as appropriate. Also, each day, Connie receives copies of purchase orders from Stan, which she uses to update quantity on order (QOO) data, adding the order quantity for each plant to the QOO. (See Figure IMS-4 for a sample purchase order.) As orders are delivered, Connie receives copies of the shipping lists (see Figure IMS-5) from the Greenhousing Department and adds the quantities received to the QIS count and subtracts them from the QOO count. She also updates the IS/OB and OO/OB quantities, shifting OO/OB quantities to IS/OB as appropriate. For example, when the order of 3 laccospadix australasica palms is received, Connie updates the counts to indicate that QIS equals 6, IS/OB equals 4, and QOO and OO/OB equal zero.

In addition to updating inventory each time an order is placed, a shipment arrives, or a designer includes an item in a bid, Connie also keeps track of the stock items removed from physical inventory. Overseeing the physical inventory is the responsibility of the Greenhousing Department, which also performs all shipping and receiving tasks as well as caring for the plants in inventory. As Production crews load plants and other materials from inventory onto their trucks each morning, Bill Johnson, the Greenhousing clerk, fills out the Materials Used portion of NBD's Daily Work Report form (see Figure IMS-6), listing the description and quantity of items removed. Bill signs the form and gives it to the Production crew. At the end of the day, each Production crew submits these forms to Connie. Connie subtracts the listed materials from her QIS and IS/OB totals.

Natural by Design, Inc.

14398 Union Road   San Jose, CA 95190

(408) 775-3600

Supplier: Greens Etc.   **Address:** 1945 E Palmetto Road
San Raphael, California

PO Number: NBD054-94   **PO Date:** 5/15/96

Bill To:
Natural by Design, Inc.
14398 Union Road
San Jose, CA 95190

Ship To:
(same)

Qnty	Description	Size	Price/Unit	Extended Price	
5	arecastum coco plumosa	15 gal	284.00	1,420.00	BG
3	arenga pinnata palm	15 gal	315.00	945.00	bo/BG
			Sub-total	2,365.00	
			Tax	189.20	
			Total	2,554.20	

Figure IMS-4: NBD Purchase Order

Greens Etc.

1945 E Palmetto Road

San Raphael, California

Invoice Number: GR0976-94   **Account:** NBD
Invoice Date: 5/28/96   **PO Number:** NBD054-94
Ship Date: 6/05/96   **PO Date:** 5/15/96

Bill To:
Natural by Design
14398 Union Road
San Jose, CA 95190

Ship To:
(same)

Qnty	Description	Size	Price/Unit	Extended Price
5	arecastum coco plumosa	15 gal	284.00	1,420.00
	Note: arenga on back order; will ship and invoice separately			
			Sub-total	1,400.00
			Tax	113.60
			Total	1,513.60

Figure IMS-5: Shipping List from Greenhousing Clerk

Production Daily Work Report

Date: 6/17/96
Project: LS Mall
Submitted by: Monica Goce

Materials Used:

Code	Qnty	Unit Cost	Ext. Cost	Approved by
cary	5	143.00	715.00	B Johnson
margi	7	45.00	315.00	B Johnson
cedar bark	10	7.50	75.00	B Johnson
Total Cost of Materials:			1,105.00	

Labor:

Worker	Hours	Cost/Hr	Ext. Cost	Task
Monica	8 hrs	18.00	144.00	installed plants and bark
Bert	8 hrs	18.00	144.00	installed plants and bark
Total Cost of Labor:			288.00	

6/17/95 — Date

Monica Goce — Lead Worker Signature

Figure IMS-6: Production Daily Work Report

E. Increasing Competition

A major concern for NBD is competition. In recent years, a number of similar firms have begun operation, and business that at one time was easy to obtain has now become very competitive. Due to competitive pressure, NBD is increasingly required to reduce its prices. While, on the one hand, Keri and Stan strive to provide the best and not necessarily the cheapest product, on the other hand, they cannot entirely ignore price levels set by their competition. To retain profitability in this environment, NBD must be as efficient as possible in all aspects of its business. This means buying at a low price, minimizing inventory costs, and passing on these savings to their customers.

F. Current Use of Information Technology

Currently, NBD has made only limited use of information technology to improve its efficiency and effectiveness. The designers have access to one microcomputer and a laser printer, which they use to prepare design bids. Connie, who is taking evening classes at the local university to complete an MIS degree, recently developed an Excel spreadsheet template to help the designers prepare design bids and to improve the format consistency and the mathematical accuracy of this form. Connie also has a microcomputer; she uses an Excel worksheet to maintain inventory count data and Microsoft Word for Windows to generate miscellaneous reports. She shares the Design Group laser printer. A third microcomputer, a second laser printer, and another Excel worksheet are used by Stan. Stan prepares purchase orders manually.

II. Inventory Management System Project

Stan and Keri want to develop a computerized Inventory Management System (IMS) to help them keep track of their inventory and to integrate the Design, Greenhousing, and Purchasing functions. They've heard about systems that use electronic forms to generate purchase orders and bids and data entry screens on which inventory and shipping/receiving personnel can record all materials received into or removed from inventory. In these systems, the inventory database is instantly and automatically updated as each transaction occurs. For example, when Stan generates a purchase order, the number of an item on order would be updated; similarly, when a designer creates a design bid, the number of an item on bid would be updated. Automatic updating of the database will not only reduce Connie's workload but also provide up-to-the-minute inventory information for designers.

Lacking their own IS staff, Stan and Keri have decided to hire a consulting team to develop the system. Your task as a member of the consulting team is to determine the requirements of an information system that will maintain up-to-the-minute inventory counts by tracking all purchase orders, shipping lists, design bids, and production work reports. You are to analyze these documents only as they affect inventory. That is, you need to sift through the myriad details in these documents to identify which data are needed to maintain inventory status and to determine when and how much stock to order. You do not need to model all the project or cost data, nor do you need to keep track of clients, project teams, accounts receivable, or accounts payable. You should assume that, if Stan and Keri want to automate these functions, they will fund future development projects.

In addition, to minimize costs, Stan and Keri have stipulated that NBD's existing hardware devices (three 486 DX2 microcomputers and two HP LaserJet printers) must be incorporated into the new system configuration and that commercially-available software must be used to construct the system. The new IMS must be operational within 6 months, and development costs must not exceed $55,000, including consulting fees, any additional hardware or software, training, documentation, data conversion, and so on.

III. Interviews with NBD Users

Given Connie Nguyen's interest in MIS, Stan and Keri asked her to be part of the consulting team and to guide the team through interviews with NBD employees. The following interviews provide more detail on how the design bid, shipping list, and production daily work report are prepared.

Interview with Tamara Bakken, Designer

The consulting team, accompanied by Connie Nguyen, met with Tamara to determine how the designers interact with the Inventory Management System. Connie began by asking Tamara how she prepares a Design Bid.

"I have to admit that, after six years as a NBD designer, I've fallen into an almost unconscious -- if that's the right word -- method of estimating materials and design and labor hours. I refer to my preliminary sketches to determine the materials needed. Estimating the total price of plants and materials is pretty clear cut. Connie distributes updated price lists for all inventory items on each day's inventory report. All we designers do is find the desired item on the list and enter it and its price on the spreadsheet template that Connie created (see desbid.xls). We enter the quantity of each item, and then the spreadsheet calculates the extended price and the total price for materials. The finished bid is approved by Cheryl and then shown to the client and, if approved, I retain one copy and give the original bid to Connie, who, I guess, forwards one copy to Sue Kaufman in Production and another to Accounts Receivable, and retains the original to file and to update inventory data on her spreadsheet. Is that right, Connie?"

"Pretty much," Connie said. "For now, we just want to understand what you do. If you can't think of any other details about how the designers prepare bids, do you have any complaints about the current system? Are there any improvements you'd like to suggest?" Connie asked.

"Hmmm, let me think. It would be nice if all I had to do to prepare the bid was enter the inventory code and quantity for materials and the system would retrieve the appropriate description, price, etc. Same with labor. If we had codes for labor type, all I'd have to do is enter the code, and the description and hourly price would be plugged in. Is that within the scope of your project?"

"I'm not sure," Connie responded for the group. "We'll have to talk with Stan and Keri about that. Do you have any other complaints or suggestions?"

"Well, currently, I feel quite constrained as a design a landscape. I have to know when implementation will likely occur so that I can select plants that will be available. I guess what I'm trying to say is that, if a design will be implemented immediately upon approval, I pretty much have to go with plants that are currently available. I sometimes find myself steering a client away from certain plants because I know that we're unlikely to have them in time for implementation."

"What would the system have to be able to do to help you avoid this constraint?" one of the consultants asked.

"I have no idea .... I guess if it could shorten the lead-time, that would be a big help."

"Can you think of anything else we should know?"

"Not that I can think of now. But I'll let you know if I come up with any other ideas."

Interview with Bill Johnson, Greenhousing Clerk

"Hi, Connie. Folks. Come on in. Sorry I don't have enough chairs for everyone, but I don't think this will take too long."

"Hi, Mr. Johnson," Connie said and then introduced the members of the consulting team. "If I understand correctly, your interactions with the inventory tracking process include the Production Daily Work Report (PDWR) and the shipping list. Let's start with the PDWR. Could you tell us about your role in completing that report."

"Well, it's pretty simple. All I do is record the unit cost and extended cost for items removed from inventory and sign my name. The lead production worker has already written the date, project name, and item code and quantity on the form when I see it. So that's about it. Told you this would be a short meeting."

"How do you know what unit cost to enter for each item?" asked a consultant.

"Pretty simple. All I do is refer to the most recent inventory report that lists the net cost and list price for all inventory items. Then I calculate the extended cost. That's it."

"Do you ever have any problems, like entering the wrong unit or extended cost?"

"Oh, occasionally, I botch the job. Especially if I'm trying to do two things at once. Some mornings are more hectic than others. But Connie here usually takes me to task for any errors I make, don't you, Connie?"

"Well, sometimes," Connie responded, "but I live in dread that I'll let errors slip by me. Can you think of any way we could avoid the occasional calculation error?"

"Humph. That's your job, not mine. Don't know much about those computer things and can't say that I care to learn. I think it's a disgrace that kids nowadays can't compute in their heads. Have to have a calculator to add two plus two. Not me. I do all my calculating by hand, thank you."

"Well, for instance, would it help if all you had to do was enter the item code and quantity and the system would retrieve the unit cost and calculate the extended cost?" one of the consultants hesitantly suggested.

"Entering the codes and quantities is the lead production worker's job. Not that I don't know all the codes, mind you, but I tend to believe the best 'system' is one where everyone does his job and doesn't stick his nose in anyone else's business."

"Okay, then, what about the shipping list? How do you process that document?" one of the consultants asked.

"Again, that's pretty simple. All I do is retrieve the appropriate purchase order -- Stan sends me a copy each time he prepares one -- and compare the items on the shipping list to those on the purchase order. Usually things match. If they don't, I call Stan to let him know about a shipment error -- you know, the supplier sent 12 arecastum coco plumosa instead of 10, or substituted 17-inch ficus green gems for 14-inch. Usually, Stan tells me to accept the shipment and forward the shipping list to him, instead of to Connie here like I usually do. I guess Stan waves a magic wand over it and then sends it on to Connie."

"A magic wand?" a consultant interrupted.

"Yeah, a magic wand. In other words, he signs his John Henry to indicate that he approved the shipment."

"Oh, I see. Do you mark the shipping list in anyway?"

"Sure thing. On the purchase order, I initial each line that agrees with our purchase order to indicate that we received that quantity of that plant or whatever. I circle each quantity that's in error and note the shipment quantity; same with items -- circle any ordered items that don't agree with the shipped items and indicate the name and quantity of the item shipped in error. 'Course, sometimes, the supplier sends something totally different from what was ordered. You know, sends laccospadix australasica palms instead of marginata. That's usually the kind of shipment error that Stan has me return, unless, of course, we happen to be low on lacco palms!"

One of the consultants frowned, scratched his head, and asked, "Would it be correct to say that you usually accept whatever the supplier sends you? From the sounds of it, you're usually so desperate for plant inventory that you'll accept whatever you can get."

"That's pretty much the case. Guess that's where you come in. You high-paid consultants are going to throw some computers at us to make the problem go away."

"Okay, Mr. Johnson, you've been very helpful. Thanks for your time," Connie said as the consultants filed out of the cramped office.

"No problem, Connie. Just don't go thinking about putting one of those computer things down here in the greenhouse. Too damp. Can't tell what might happen."

Interview with Monica Goce, Lead Production Worker

"Hi, Monica," Connie said. "Thanks for agreeing to meet us after work. Let me introduce the consulting team." After introducing the team, Connie turned the interview over to a consultant.

"We've already talked to Bill Johnson about his role in completing the Production Daily Work Report. Now we need your take on the subject."

"That old goat! He fries my pajamas. Just this morning we had a big to-do because he had entered the wrong extended cost for an item. Do you think he could admit that he had multiplied wrong? No way!" Monica shook her head in disgust.

"Does that kind of thing happen very often?"

"More often than it should, or would if he'd just condescend to use the calculator sitting on his work table. "

"What do you do when you know that a figure's wrong?"

Monica chuckled. "I just wait until he's out of sight, then I scratch out his error and enter the right number. There's no sense trying to reason with the man."

"I think we're getting ahead of ourselves," Connie said, trying to change the topic away from employee politics. "Monica, would you explain how you complete the PDWR?"

"Sure, sorry for the digression. There's a stack of PDWR forms just inside the Greenhousing office. The first thing I do each morning is to consult the Production Plan (see Figure IMS-7) folder to determine what project or projects my team is working on that day and what inventory items we need to take to each site. Then I enter the date, project code, item code, and quantity on a form for each project. Usually doesn't take me more than 30-40 minutes. Then I find our driver and have him bring the truck around to the loading dock. I hand the forms to old Bill and keep an eye on him to make sure that he gets all the right stuff loaded on the truck. At the end of the day, I write in the names of the employees who worked on each project, the number of hours, and a description of the work done. I drop the forms in Connie's in-box before I punch out."

"What about the cost per hour and extended cost figures?" one of the students asked.

"Oh, yeah. I enter the average cost per hour for a particular kind of employee -- for production workers, that's about $18 an hour -- and multiply by the number of hours."

"Do you ever make mistakes?"

"Seldom. Unlike some people I could name, I'm not too proud to use a calculator. I usually perform the calculations manually and then verify them with a calculator."

"Can you think of any changes that would make your job easier?"

PRODUCTION PLAN
June 1, 1995

Project Bid Information

Project:	LS Mall	Project Site:	Main entrance
Begin Date:	June 14, 1995		Mall Dr./Cinema Lane
Compl. Date:	June 18, 1995	Bid Amount:	$7,457.50

Project Team

Sales Assoc:	Bob Reinhardt	Production:	Sue Kaufman, Monica Goce,
Designer:	Tamara Bakken		Bert Swenson

Material Requirements

Qnty	Code/Desc	Size	Net/Unit	Ext. Cost	Deliver	Install
Plants:						
3	lacco	15 gal	450.00	1,350.00	6/16 @ 8 am	6/16
5	cary	7 gal	140.00	700.00	6/17 @ 8 am	6/17
7	margi	2 gal	45.00	315.00	6/17 @ 8 am	6/17
Pottery:						
1	GFN48	48 in	457.50	457.50	6/15 @ 8 am	6/15
3	GP50	50 gal	110.00	330.00	6/15 @ 8 am	6/15
Materials:						
10	CBRK5	5 cu ft bag	7.50	75.00	6/17 @ 8 am	6/17
1	TSOIL	yard	12.50	12.50	6/15 @ 1 pm	6/15
Tools:						
2	contouring rakes				6/15	
2	spades				6/16 - 6/17	
2	landscaping rakes				6/16 - 6/17	

Labor Requirements

Desc	Hours	Cost/Hr	Ext. Cost	Time	Tasks
Design consultant	12	40.00	480.00		Bid Process
Tamara	5	40.00	200.00	by 6/10	Complete final blueprint
	3	40.00	120.00	6/15 10 am	Oversee installation of fountain & pots
	1	40.00	40.00	6/16 8 am	Inspect contouring
	1	40.00	40.00	6/18	Inspect finished site
Production workers					
Monica & Bert	6	18.00	108.00	6/15 2 pm	Contour surface
Monica & Bert	8	18.00	144.00	6/16 8 am	Install large plants
Monica & Bert	16	18.00	288.00	6/17 8 am	Install small plants & bark
Eqpmt operator	6	45.00	270.00	6/14	Remove existing structure
Jerry	4	45.00	180.00	6/15	Situate fountain & pots

Total Est. Designer Hours/Costs:	**22**	**880.00**	
Total Est. Labor Hours/Costs:	**40**	**990.00**	
Total Est. Material Costs:		**3,240.00**	
Total Est. Hours/Costs:	**62**	**5,110.00**	**(67 percent of bid)**

Figure IMS-7: Production Plan

"Sure, if talking the old goat into retiring is something you can include in your system design! Just kidding. He's really not as bad as I make out. My job wouldn't be half as challenging if I didn't have to fight Billy every step of the way. There I go again. Digress. Digress. Digress. The change I'd most like to see is for the system you're working on to automatically fill in as much as possible on the PDW Report. I mean, all I'm doing is copying information from the Production Plan."

"So, you're saying that the item code and quantity for each project for each day is already listed on the Production Plan. All you do is copy it?"

"That's it. Seems kind of silly. The worker names, tasks, and estimated hours are also on the Production Plan. So it seems that your system could generate a default PDW report and I could just edit it to note any changes to plan."

"Are there often changes to plan?"

"Runs in streaks. For days on end, everything goes according to plan. Then all of a sudden we'll have unexpected stockouts, or a job will take much more or much less time than expected, or a worker won't show up for a couple of days. Or sometimes a designer will inspect a job in progress and decide that we need to install eight marginata instead of the seven in the blueprint. Designers are given some discretion to make goodwill kinds of adjustments at no cost to the client. They don't go through the bother of revising the bid and plan; they just make the change, getting Stan's or Keri's approval if the cost is significant."

"Are the changes common enough that generating a default form wouldn't be practical?

"No, I'd guess that at least 95 percent of the time I'm just copying the production plan word-for-word. So, yeah, if you could generate a default PDWR and let me edit it if necessary, that would be helpful."

"Using a default PDWR data entry screen would also help you complete the form more easily and accurately. You'd enter the project code and date and then either accept the defaults or edit the item codes and quantities. Then the unit cost and extended cost would be computed by the system. Same with the labor data; you'd enter the worker name, labor code, number of hours, and task, and the system would do the rest."

"I don't know. I usually enter materials right away in the morning and labor data at the end of the day. Could I do that with the screen you're talking about?"

"Sure. You could save the morning data, bring up the incomplete screen at the end of the day, and finish entering data."

"But how would Bill approve the items removed from inventory and how would I sign the form?"

"Oh, I'm sure we could figure something out. Anyway, the main thing is, would such a data entry screen make your job easier and improve the accuracy of the PDW reports?"

"I 'spose so. I'd have to see it in action before I could say for sure."

Interview with Stan Felton and Keri Yamaguchi

Connie and the consulting team met with Stan Felton to learn more about the purchasing process and his expectations for the system. Keri Yamaguchi was supervising an exceptionally high-priced project and promised to join them as soon as she could get away.

"Formerly, we were incurring unreasonably high procurement and inventory costs because we had no formal process for determining the optimal order and inventory quantities for each item. But, thanks to Connie, I now use a worksheet decision model to help me determine the economic order quantity and economic reorder point for each plant type (see Figure IMS-8). Perhaps Connie can explain this for you."

"Sure, Stan. The economic order quantity (EOQ) is the quantity of a stock item that a firm should order at one time to minimize its costs of both procurement -- the process of ordering and receiving goods -- and storage and to reduce the amount of capital it has tied up in inventory.[1] The economic reorder point (ERP) indicates the stock level at which an item should be reordered to avoid stock-outs.[2] By determining the EOQ and ERP for each plant type, we can more effectively manage our inventory and reduce the costs of purchasing and storing our stock."

Connie continued, "The procurement costs and inventory cost rate[3] (shown in the worksheet) must be known in order to determine the appropriate order and inventory quantities for each plant type. Also needed are the average net cost and annual sales for each plant type (shown in Figures N-2 and N-3)."

---

[1] The formula for calculating the EOQ is

SQRT((2 * procurement cost)/inventory cost rate) *
SQRT(annual unit sales/unit cost)

where *SQRT* indicates the square root.

[2] The formula for calculating the ERP is

order lead time * weekly unit sales

where *order lead time* is the number of weeks from the date of placing an order to the date of receiving it and *weekly unit sales* is the annual unit sales divided by 52.

[3] Inventory Cost Rate is represented as a percentage of the plant type's net cost. For example, if a plant type's net cost is $100 and the inventory cost rate is 42 percent, its inventory cost is $42.00.

INVENTORY MANAGEMENT DECISION MODEL

File name: EOQ-ERP.XLS

Plant Code	Net Cost	Annual Sales	Lead Time (wks)	EOQ	Safety Stock	ERP w/SS	ERP no SS
lacco	$450.00	100	4	7	2.75	10	8
arenga	$310.00	160	2	11	2.46	9	6
cham	$300.00	200	4	12	3.88	19	15
cera	$240.00	175	4	13	3.63	17	13
areca	$275.00	265	3	14	3.87	19	15
cary	$140.00	350	3	23	4.45	25	20
phoenix	$105.00	300	5	25	5.32	34	29
bamboo	$110.00	225	8	21	5.82	40	35
raphis14	$350.00	275	4	13	4.55	26	21
spath	$90.00	350	3	29	4.45	25	20
cory	$130.00	500	3	29	5.32	34	29
ti char	$155.00	250	7	19	5.74	39	34
red ti	$140.00	300	7	22	6.29	47	40
grti7	$140.00	400	5	25	6.14	45	38
jc	$55.00	600	2	49	4.76	28	23
grti5	$92.00	600	5	38	7.52	65	58
raphis3	$180.00	300	4	19	4.76	28	23
ficus14	$90.00	500	4	35	6.14	45	38
ficus17	$240.00	350	4	18	5.14	32	27
margi	$45.00	500	2	49	7.24	26	19

INV COST RATE:	
Maintainance	20%
Depreciation	8%
Insurance/Tax	5%
Interest	9%
Inv Rate	42%

PROCUREMENT COSTS:	
Communications:	$5.00
Labor:	$35.00
Supplies:	$2.50
Miscellaneous:	$3.00
Total	$45.50

Safety Level: 95%

Safety Level	Multi-plier
10%	-1.28
20%	-0.84
30%	-0.52
40%	-0.25
50%	0
60%	0.25
70%	0.52
80%	0.84
90%	1.28
95%	1.65
98%	2.05
99%	2.33

Figure IMS-8: EOQ/ERP Worksheet

"What you should notice here," Stan explained, "is that our inventory cost rate is exceptionally high, largely because of the high cost of maintenance. Someone has to take care of the plants in inventory on a daily basis -- water, fertilize periodically, remove dead leaves, inspect for disease, whatever. We keep a full-time botanist on staff not only to recommend plants for certain settings but also to inspect and treat our inventory of plants. And, of course, any plants that 'die' in our care are a total loss unless we can trace the disease back to the supplier, in which case the supplier may agree to replace the diseased plant."

"Your procurement costs also seem rather high," one of the consultants noted. "I imagine the labor cost is primarily the cost of your time to review inventory levels, consult the worksheet, and then complete a purchase order."

"That's right," Stan responded. "If we could reduce procurement costs, the EOQ would be lower. Thus, we could order fewer plants more often, in turn reducing our maintenance costs. What I'd like you to do is to set up the system so that each plant type has an assigned reorder point. When the quantity in stock is equal to or less than the reorder point, the system should automatically alert me to place an order for that plant."

"That's quite feasible," a consultant explained. "In fact, that's the key to reducing your costs. In some ways, what you need is a just-in-time inventory system. If a plant is received from the supplier within 24 hours of the time it is to be installed in a landscape project, your inventory maintenance costs would be minimized."

"Yes, but, as you can see, most of our suppliers require at least a 2-week lead time."

"Right," the consultant agreed. "So the trick is to figure out how to reduce that lead time. That goal should be feasible, especially with local suppliers. I mean, you're in California, for Pete's sake! There must be a number of nurseries that could guarantee over-night delivery."

"That's true ...." Stan paused in thought. "Maybe my real problem is that I haven't been demanding the level of service I need. I'm a pretty easy-going guy; I tend to take what I'm given. When we were just starting out, that was necessary. But we've grown into a pretty large operation, so I should have some leverage with my suppliers. I mean, I'm doing in excess of $100,000 a year with many of them."

"From what we've heard, though, some of your suppliers aren't very dependable."

"What gave you that idea?" Keri interrupted, joining the group. "Sorry to be late. Unfortunately -- or perhaps I should say 'fortunately' -- a mega-bucks project takes precedence! Anyway, our suppliers would not be 'preferred' if they weren't dependable."

"Thanks for making time for us," a consultant said and then explained, "Bill Johnson noted problems with receiving unordered items or receiving the right item but the wrong quantity."

"Bill. He's just never been able to forget the time we placed an emergency order with a new supplier who messed up the order and cost us a project. That was a once-in-a-blue-moon incident that Bill insists on blowing way out of proportion," Keri laughed. "If you learn anything during this interview process, you should learn that Bill's view of things should not be taken too much to heart!"

"Then do you think it's feasible to reduce lead times?," Stan asked. "That's what we were discussing when you came in, Keri."

"I'm pretty sure of it. In fact, I've recently begun hinting to several of our suppliers that we can't continue to operate competitively if they have a strangle-hold on the availability of our plant stock."

"Great," a consultant concluded. "Then, we'll pursue this further with you, Keri. For now, let's talk more generally about your expectations for the new system. What business goals do you expect the system to achieve?"

Stan replied, "We've just enumerated a number of them in our discussion: reducing procurement and inventory costs, reducing supplier lead times, making us more competitive -- if our costs are lower, we can offer our clients lower prices. Overall, we simply need to be able to provide better service to our clients. Clients want a wide choice of plants, pottery, and materials; many of them also want immediate implementation."

"How immediate is 'immediate'?"

"Some clients plan ahead and get the process moving as much as six months before they expect implementation to begin. A few expect us to begin implementing as soon as the ink is dry on their bid signature. In general, we need to begin implementing within four -- at most, six -- weeks of bid approval. Longer than that and we risk losing the bid."

"Absolutely," Keri agreed. "We've lost some projects because we couldn't implement the landscape as quickly as the client wanted. We had innovative designs, a ready labor force, but we couldn't get the desired plants quickly enough."

"I'm reminded of my mother's favorite quotation," one of the consultants mused. "Something to the effect that 'For lack of a nail, the horse was lost; for lack of a horse ....' I don't know, basically the kingdom crumbled. Anyway, your situation seems to be 'for lack of a plant, the project was lost.'"

"You've hit the 'nail' on the head," Keri responded. "Our whole business hinges on the availability of plants. We've worked out all the other details: creative, personable designers who create fabulous designs and who interact famously with clients; hard-working, dependable production workers who do a quality job implementing the designs; and so on. But there just isn't any landscape without plants."

"Amen," a consultant said. "On that note, we'll let you get back to work. And we'll get to work 'saving the kingdom.'"

Interview with Connie Nguyen

"OK, Connie. It's time for you to submit to our interrogation."

"Please be gentle," Connie joked.

"Yeah, right," one of the consultants said. "What we most need from you at this point is an explanation of how plant costs and prices are calculated. We understand that these tend to be more volatile than pottery or material prices. Stan said something about your maintaining the average cost and computing the selling price from that. Could you explain, please?"

"Sure, it's pretty simple. Each shipping list indicates the quantity, unit cost, and extended cost of the plants received. Whenever I receive a shipping list, I add the quantity of each plant to the total annual quantity of that plant; I also add its extended cost to the total annual cost of that plant. Then I divide the total annual cost by the total annual quantity, giving the average cost. To calculate the selling price, I multiply the average cost by 1.665, which means that we have about a 67 percent mark up on plants."

"Isn't that a little high?" one of the consultants asked.

"No, in fact, it may be a bit low. You have to remember that the labor costs of preparing the design bid are factored into the selling price of plants. The only time we charge up-front for pre-bid-approval design labor is when the project is exceptionally complicated, thus requiring a significant investment."

"Okay, I understand the rationale. So, for example, if we look at the inventory report (see Figure IMS-2), the average cost (AveNet$) for lacco is $450; multiplying that by 1.665 gives a list price (List$) of $749. Currently, you do this manually, right?"

"That's partially true. I compute the totals manually, but I use the Excel spreadsheet (see nbd-inv.xls, specifically the YTD-Qty and YTD-Cost columns) to calculate the average cost and the list price."

"Okay, that's pretty clear. Just one more question. We have a pretty clear idea of how you maintain the QIS, IS/OB, QOO, OO/OB counts, tedious as it must be for you."

"Truer words were never spoken!" Connie interrupted. "Sometimes I feel completely bonkers by the time I finish all those machinations."

"I can believe that," the consultant responded. "So you must go totally off the deep end when you receive a bid that lists plants whose in-stock and on-order quantities are already on bid. In other words, all in-stock and on-order plants are already spoken for. What do you do then? There's no place on your spreadsheet to deal with on-bid but neither in-stock nor on-order quantities."

"Fortunately, that seldom happens because designers know that the likelihood of actually being able to use those plants is almost nil, unless, of course, the project won't be implemented for a few months. Anyway, when that does happen, I run to Stan, begging him to place an order immediately, if for no better reason than so that my spreadsheet will make sense. But Stan's pretty sharp. He's probably already noticed that all current and on-order stock is on-bid, so he's probably in the process of ordering or just hasn't forwarded his order to me yet."

"Good," the consultant summed up. "Let's hope the new system saves both you and Stan some of these headaches."

"Hope?" Connie mocked. "You'd better do more than hope if you plan to earn your consulting fees!"

A PROJECT ADMINISTRATION SYSTEM FOR *NATURAL BY DESIGN*

I. Background

Natural By Design (NBD) is a partnership that designs and develops interior and exterior landscapes for hotels, restaurants, condominiums, and large private estates. In their designs, they use a mixture of commonly available plants, flowers, and shrubs, and imported rare and exotic plants, pottery, and other materials. *NBD* employs twenty-five people and has a gross sales of $4.1 million with a before-tax profit of $750,000. Keri Yamaguchi and Stan Fenton own the partnership equally. Keri, a designer, manages the artistic side of the business -- including not only supervising the design activity, but also job production and the selection of vendors. Stan is the general manager who manages the inventory and supervises the sales and financial activities.

A. Design Process

NBD's Design Group employs six full-time designers; an administrative assistant, Connie Nguyen; a botanist, and a group manager, Cheryl Poy. Cheryl works closely with Keri, who remains the firm's chief designer. This arrangement could be problematic except that Cheryl and Keri have worked together for many years, and they have established a relationship in which Cheryl manages the more administrative concerns of the design group while Keri oversees and contributes to the major design activities.

Each designer potentially has many projects that s/he works on, and each project potentially has many designers. After one of *NBD*'s salespeople has identified a real prospect, the designers meet with clients to determine the client's requirements. Depending on the size and complexity of a project, the designers may produce one or more preliminary sketches. They also prepare a formal Design Bid (see Figure PAS-1) that contains a general description of the project, a list of plants and materials, and a detailed list of both material and labor prices, including the price for any designer hours required after the bid has been approved.

Natural By Design

14398 Union Road   San Jose, CA 95190

(408) 775-3600

Design Bid

Client

Client Name:	London Sq. Mall	Contact:	Amy Benson, Mgr.
Client Address:	12638 Mall Drive	Phone:	(408) 834-5603
	Scotts Valley, CA 95060		

NBD Staff

Sales Assoc:	Bob Reinhardt	Phone:	(408) 775-3652
Designer:	Tamara Bakken	Phone:	(408) 775-3645

Project

Bid Date:	May 6, 1996	Project Site:	Main entrance
Est. Begin Date:	June 15, 1996		Mall Dr./Cinema Lane
Est. Compl. Date:	June 30, 1996	Bid Amount:	$7,651.50

Material Requirements

Plants:

Qnty	Desc	Size	Unit Price	Extended Price
3	laccospadix australasica palm	15 gal	$749.00	2,247.00
5	caryota mitis	7 gal	$233.00	1,165.00
7	marginata	2 gal	$75.00	525.00

Pottery:

1	granite fountain (GFN48)	48 in	$750.00	750.00
3	granite pots (GP50)	50 gal	$195.00	585.00

Materials:

10	decorative cedar bark (CBRK5)	5 cu ft	$15.95	159.50
1	top soil (TSOIL)	yard	$20.00	20.00

Labor Requirements

Hours	Desc	Unit Price	Extended Price
30	Production workers	$30.00	900.00
10	Design consultant	$65.00	650.00
10	Heavy equipment operator	$65.00	650.00

Figure PAS-1: Design Bid

NBD bills a client for pre-bid-approval design time only if the project is exceptionally large or risky; in all other cases, the costs of preparing the Design Bid are factored into *NBD*'s markup for plants, materials, and production-phase labor. Cheryl, the Design Group manager, must approve the Design Bid before it is shown to the client.

After the client has approved the bid, the designers develop formal sketches and blueprints, and Keri forms a team to implement the project. This team usually consists of the salesperson who initiated contact with the client, one or more designers, and several members from Production. In a large project, the Design and Production Group managers as well as Keri herself are members of the team. Because teams cut across the horizontal organization, there are no rigid, formal lines of authority. For example, the project team assigned to the London Square Mall project includes a sales representative, Bob Reinhardt; a designer, Tamara Bakken; the Production Group manager, Sue Kaufman; and the production workers, Monica Goce and Bert Swenson.

After the design has been approved and the final sketches and blueprints drawn, the designer's role is primarily that of a project overseer. For more complicated projects, the designer regularly inspects the progress of the Production crew to ensure that the design is being implemented on schedule and to specifications. The designer may also need to meet with the client if changes to the approved bid are requested.

B. Production Process

The Production Group implements the landscape designs. This group consists of the manager, Sue Kaufman; three gardeners; three laborers; and two drivers. The personnel in this group are assisted by outside specialists (for example, heavy equipment operators) and by part-time contract laborers. Production's work requires trucks, equipment, tools, supplies, materials, and plants. The vehicles, equipment, and tools must be moved from site to site.

With many projects under way simultaneously, the scheduling of workers, equipment, and tools is critical. Sue is responsible for using the designer's bid and blueprint to generate a Production Plan (see Figure PAS-2). Data provided in the Production Plan includes information on the scheduling of labor and material delivery, project task completion dates, and other implementation data. Sue submits each Production Plan to Keri for approval before the plan is implemented. Keri returns the approved Production Plan to Sue, who makes a copy and submits it to Connie for filing and tracking. If a client requests changes to the design after production has begun, the designer prepares a revised bid and submits it to Sue Kaufman. Sue augments the revised bid with production data to generate a revised Production Plan and sends the revised plan to Keri for approval.

PRODUCTION PLAN
June 1, 1996

Project Bid Information

Project:	LS Mall	Project Site:	Main entrance
Begin Date:	June 14, 1996		Mall Dr./Cinema Lane
Compl. Date:	June 18, 1996	Bid Amount:	$7,651.50

Project Team

Sales Assoc:	Bob Reinhardt	Production:	Sue Kaufman, Monica Goce,
Designer:	Tamara Bakken		Bert Swenson

Material Requirements

Qnty	Code/Desc	Size	Net/Unit	Ext. Cost	Deliver	Install
Plants:						
3	lacco	15 gal	450.00	1,350.00	6/16 @ 8 am	6/16
5	cary	7 gal	140.00	700.00	6/17 @ 8 am	6/17
7	margi	2 gal	45.00	315.00	6/17 @ 8 am	6/17
Pottery:						
1	GFN48	48 in	457.50	457.50	6/15 @ 8 am	6/15
3	GP50	50 gal	110.00	330.00	6/15 @ 8 am	6/15
Materials:						
10	CBRK5	5 cu ft bag	7.50	75.00	6/17 @ 8 am	6/17
1	TSOIL	yard	12.50	12.50	6/15 @ 1 pm	6/15
Tools:						
2	contouring rakes				6/15	
2	spades				6/16 - 6/17	
2	landscaping rakes				6/16 - 6/17	

Labor Requirements

Desc	Hours	Cost/Hr	Ext. Cost	Time	Tasks
Design consultant	12	40.00	480.00		Bid Process
Tamara	5	40.00	200.00	by 6/10	Complete final blueprint
	3	40.00	120.00	6/15 10 am	Oversee installation of fountain & pots
	1	40.00	40.00	6/16 8 am	Inspect contouring
	1	40.00	40.00	6/18	Inspect finished site
Production workers					
Monica & Bert	6	18.00	108.00	6/15 2 pm	Contour surface
Monica & Bert	8	18.00	144.00	6/16 8 am	Install large plants
Monica & Bert	16	18.00	288.00	6/17 8 am	Install small plants & bark
Eqpmt operator	6	45.00	270.00	6/14	Remove existing structure
Jerry	4	45.00	180.00	6/15	Situate fountain & pots; shape surface

Total Est. Designer Hours/Costs:			
Bid Process	**12**	**$480.00**	
Production	**10**	**$400.00**	
Total Est. Prod. Labor Hours/Costs:	**40**	**$990.00**	
Total Est. Material Costs:		**$3,240.00**	
Total Est. Hours/Costs:	**62**	**$5,110.00**	**(67 percent of bid)**

Figure PAS-2: Production Plan

C. Tracking Project Costs

Except for very small jobs, each designer (or the lead designer in the case of a larger project requiring two or more designers) is required to submit a budget of the time s/he expects to devote to preparing the bid (see sample Design Budget in Figure PAS-3). The Design Budget is submitted to Keri for approval after a preliminary meeting with a client so that the designer can realistically estimate the hours required to design each project.

NBD has a general policy that its designers may invest approximately 10 percent of the potential project's bid amount in design activities; thus, as illustrated by the Design Budget in Figure PAS-3, for a project with an estimated bid of $7,500, the maximum design hours is 19 ($40 per designer hour * 19 = $760, or approximately 10 percent of the bid amount). Keri and Stan instituted this policy because designers tended to get carried away and to invest more design time in a bid than management deemed appropriate.

Design Budget

Client

Client Name:	London Sq. Mall	Contact:	Amy Benson, Mgr.
Client Address:	12638 Mall Drive	Phone:	(408) 834-5603
	Scotts Valley, CA 95060	Contact Date:	April 15, 1994

***NBD* Staff**

Sales Assoc:	Bob Reinhardt	Phone:	(408) 775-3652
Designer:	Tamara Bakken	Phone:	(408) 775-3645

Project

Budget Submitted:	April 19, 1996	Project Site:	Main entrance
Est. Begin Date:	June 15, 1996		Mall Dr./Cinema Lane
Est. Compl. Date:	June 30, 1996	Bid Amount:	@ $7,500

Design Budget

	Task	**Date**	**Hours**
Hours To Date :			
	Initial mtg w client	April 15	1
Estimated Hours:			
	Prepare preliminary sketches	April 21	2
	Prepare design bid	April 21	2
	Meet with client	April 25	1
	Revise sketches & bid	@April 25	2
	Meet with client	@ May 2	1
	Prepare initial blueprint	@ May 3	3
Total Hours:			12

Tamara Bakken — Submitted by

April 15, 1996 — Date

Figure PAS-3: Design Budget

NBD uses two forms to track the actual costs of materials and labor for each project. Production workers submit a Production Daily Work Report (see Figure PAS-4), indicating the material and labor expended on each project each day. Notice that as a control measure, Bill Johnson, the Greenhousing clerk, must indicate the cost of all plants, pots, and materials and must approve their removal from inventory. In addition, each designer submits a Designer Daily Work Report indicating the hours devoted to each of his or her projects (see Figure PAS-5).

Copies of all forms -- Design Bid, Production Plan, Design Budget, Production Daily Work Report, and Designer Daily Work Report -- are submitted to Connie Nguyen, the Design Group administrative assistant on a daily basis. At the start of every business day, Connie records the bid amount from new or revised Design Bids, the estimated production costs for labor and materials from new or revised Production Plans, and the actual costs of material and labor for each project from Production Daily Work Reports and Designer Daily Work Reports. Each Monday morning, Connie summarizes this data in a report showing the estimated and actual costs to date for each project (see Figure PAS-6) and delivers it to the Stan, Keri, the Group managers, and the designers.

Production Daily Work Report

Date: 6/17/96
Project: LS Mall

Materials Used:

Code	**Qnty**	**Unit Cost**	**Ext. Cost**	**Approved by**
cary	5	143.00	715.00	B Johnson
margi	7	45.00	315.00	B Johnson
cedar bark	10	7.50	75.00	B Johnson
Total Cost of Materials:			1,105.00	

Labor:

Worker	**Hours**	**Cost/Hr**	**Ext. Cost**	**Task**
Monica	8 hrs	18.00	144.00	installed plants and bark
Bert	8 hrs	18.00	144.00	installed plants and bark
Total Cost of Labor:			288.00	

6/17/96
Date

Monica Goce
Lead Worker Signature

Figure PAS-4: Production Daily Work Report

Designer Daily Work Report

Date: 6/15/96
Submitted by: Tamara Bakken

Project	Stage	Hours	Task
SJSU	B	2	Prepared final sketches of arboretum
LS Mall	P	3	Monitored installation of fountain and pots
Fremont	B	3	Worked with client and surveyor to determine layout for formal garden and maze

6/15/96 — Date

Tamara Bakken — Designer Signature

Figure PAS-5: Designer's Daily Work Report

Project Administration Report for June 18, 1996

Bid Stage Projects

Project	Est. Bid	Actual/Est Design Hrs	Actual/Est Design Costs	Hrs/$ Remaining
Astor	$13,000	6/20	$240/$800	14/$660
Fremont	$16,250	9/25	$360/$1,000	16/$640
SJSU	$5000	9/8	$360/$320	(1)/($40)

Production Stage Projects

Project	Bid $	Prod Plan Est.	Total Cost to date	Actual Mtl $	Est. Mtl $	Actual Labor $ Prod	Est Labor $ Prod	Actual Labor $ Design	Est. Labor $ Design
LS Mall	$7,651.5	$5,110	$5,265	$3,255	$3,240	$1,008	$990	$880	$880
IBM	$11,500	$6,290	$5,705	$3,525	$3,850	$1,260	$1,440	$920	$1,000

Figure PAS-6: Excerpt from the Project Administration Report

Notice that the Production Plan, Production Daily Work Report, Designer Daily Work Report, and Project Administration Report track *NBD*'s actual direct costs for a project. For example, the actual cost per hour of a laborer or designer and the net cost of materials and plants are used to compute and track how much *NBD* has actually spent on a project. In general, *NBD* expects that direct costs will consume between 50 and 70 percent of the amount bid for each project. The London Square Mall project documented here falls within that range; the direct costs of the project ($5,110) constitute approximately 67 percent of the project bid ($7,651.50).

NBD needs to track actual costs because, once the bid has been approved, *NBD* is committed to complete the project for the amount bid. Thus, if the net cost of plants and materials increases or the designers or laborers invest more time than expected, *NBD* must absorb the added costs. Clients are responsible for extra charges only if they change the design after the bid has been approved.

II. Project Administration System

Currently, *NBD* has made only limited use of information technology to improve its efficiency and effectiveness. The designers have access to one microcomputer and a laser printer, which they use to prepare design bids. Connie, who is taking evening classes at the local university to complete an MIS degree, recently developed an Excel spreadsheet template to help the designers prepare design bids and to improve the format consistency and the mathematical accuracy of this form. Connie also has a microcomputer; she uses an Excel spreadsheet to maintain data about bid amounts and estimated and actual costs of each project; she uses Microsoft Word for Windows to generate reports. Connie would like to develop an Excel spreadsheet to generate the Project Administration report, but hasn't found the time. She shares the Design Group laser printer. A third microcomputer, a second laser printer, and another Excel template are used by Sue Kaufman, the Production Group manager, to produce a Production Plan.

One particularly stressful week, Connie worked overtime three evenings and a Saturday morning to enter all the project data. As she delivered her weekly Project Administration Report to Stan Monday morning, Connie commented on her increasing workload. "Do you have a minute, Stan?"

"Sure, Connie. Have a seat," Stan offered, clearing a stack of papers from a chair for Connie. "What can I do for you?"

"Stan, I'm going crazy. I know it's not your problem that I've chosen to take classes a couple nights a week to complete my degree. But I'm finding it increasingly difficult to give my classes the attention they require because I so often have to work overtime to complete all my duties here at *NBD*."

"I've noticed you've been staying late pretty often. In fact, weren't you here late last Tuesday? I thought you had a class Tuesday nights."

"I do, but I felt I had to miss the first hour so that I'd have those reports you asked for ready by noon the next day."

"Sorry about that. You need to learn to say 'No' or to tell me to take a hike, Connie. You're an excellent worker, but I don't want you to jeopardize your degree on our account. But, I hope you'll understand my pleasure in telling you that I don't expect things to change. This has been a banner year for *NBD*, and I expect our sales to continue their upward trend. Unfortunately, that means more bids, plans, reports, etc. for you. Do you think we should hire another assistant for the Design Group?"

"I'm not sure, at this point, that the workload justifies another body. Perhaps a part-timer." Seeing Stan grimace, Connie added, "I know we've had bad luck with part-time staff, but something has to give."

"I understand your situation, Connie, but, as you said, we don't seem to get the commitment from part-timers that *NBD* requires. And, by the time we pay benefits, profit-sharing, and a decent salary to attract a reliable part-time worker, we end up spending almost as much as we would for a full-time employee."

"Not to be nosy, but what would a part-time worker cost *NBD* each year?"

"Well, Connie, I hardly think I'd be betraying a trade secret by telling you that, with insurance, worker comp, unemployment, social security, etc., you cost about $50,000 a year. A part-timer would be closer to $30 - 35,000."

"Then maybe there's a good alternative to hiring another employee. You know, I'm studying management information systems. I'm no expert -- not yet, anyway -- but I bet it would cost you less in the long term to solve this problem by improving our information systems. We could automate the most time-consuming and repetitive part of my job, all the tedious data entry. Or redistribute it so that instead of designers and production workers filling out paper forms, they'd enter the data about costs and hours on a data entry screen. That way, I wouldn't have to reenter all their data. Not only would that save time but it would also reduce errors."

"True, but it would probably cost a small fortune to implement such a system."

"Not necessarily. We already have some of the hardware required. And the system isn't so large or complex that we'd need a heavy-duty consultant. In fact, we could get a lot of the work done at no cost."

"Connie, I admire your enthusiasm, but, believe me, there's no such thing as 'no cost.'"

"Maybe, maybe not. My systems development instructor is looking for projects for student groups to complete. It might take longer than if we hired a consultant, and a student group might not be able to do all the work. But we could determine the requirements, do the initial design, and implement a prototype or at least the simpler pieces of a production system. And the instructor supervises all the students' work, so you wouldn't have to worry about wasting a lot of time and ending up with a bum system. What do you say?"

"I say I need to talk with your instructor and with other businesses that have sponsored student projects. Then we'll see."

After consulting with Keri and contacting several local businesses that had been involved in student projects, Stan decided to approve Connie's idea. However, in talking with Connie's instructor, Stan set several conditions for *NBD*'s involvement: (1) Connie must serve as the project leader and keep *NBD* administration informed of the team's progress; (2) the students must confine their investigation to *NBD*'s cost tracking processes; (3) any system proposal must incorporate *NBD*'s existing hardware, must not exceed $40,000 for the purchase of hardware and software and for the construction and installation of the complete system; and (4) it should be feasible to implement the proposed system within 2-3 months after the initial analysis and design have been completed.

III. Interviews with *NBD* Users

Interview with Tamara Bakken, Designer

The student project group, led by Connie Nguyen, met with Tamara to determine how the designers would interact with the Project Administration System. Connie began by asking Tamara how she prepares a Design Bid (see Figure PAS-1).

"For your purposes, I guess the most relevant information concerns how I determine the bid amount. Estimating the total price of plants and materials is pretty clear cut. Connie distributes an updated price list for all inventory items with each day's inventory report (see Figure PAS-7). All I do is find the desired item on the list and enter it and its price on the spreadsheet template that Connie created (see desbid.xls). I enter the quantity of each item, and then the spreadsheet calculates the extended price and the total price for materials."

"What do you do if the quantity or type of materials changes later?" one of the students asked.

"We just bring up the appropriate spreadsheet, plug in new numbers or data, and resubmit the bid for client approval."

"Okay, let's get back to the estimation of bid amount," another student said. "How do you estimate the labor hours?"

"That's where experience pays off. I had a lot of trouble estimating the labor hours to implement a project when I first began, but, after seeing so many designs implemented, I've become pretty good at it. If I need help estimating hours for an unusual project, I talk with Cheryl or Keri or Sue, who can usually point me to an earlier project that was similar in nature. Then I just retrieve the bid or plan or project administration report for that project and use the labor hours from these to estimate my project's hours. Once I have an estimate of the hours required for each task, I consult a list of labor prices -- this one prepared by Administration (see Figure PAS-8) -- and again just plug the hourly price and number of hours into the spreadsheet. When I'm finished, the spreadsheet calculates the total labor price and the total bid amount. The finished bid is shown to the client and, if approved, I retain one copy and give the original bid to Connie, who, I guess, forwards one copy to Sue and another to Accounts Receivable, and retains the original to file and to enter data on her spreadsheet."

"Okay," Connie said. "If there's nothing more about the bid process, let's move on to the Design Budget (see Figure PAS-3) and the Designer Daily Work Report (see Figure PAS-5)."

NBD Inventory Report **June 13, 1996**

PLANT INVENTORY

Code	Desc	Size	Last Ordered	AveNet$	List$	QIS	IS/OB	QOO	OO/OB
lacco	lacco australasica	15 gal	06/11/96	450.00	749.00	3	3	3	1
arenga	arenga pinnata	15 gal	05/15/96	310.00	516.00	4	2	3	0
cham	chamaedorea	15 gal	06/09/96	300.00	499.00	0	0	2	0
cera	ceratozamia molongo	14 in	05/30/96	240.00	400.00	4	2	6	0
areca	arecastum coco	15 gal	05/15/96	275.00	458.00	11	5	0	0
cary	caryota mitis	7 gal	05/30/96	140.00	233.00	14	9	4	0
grnti5	green ti	5 gal	05/30/96	92.00	154.00	11	8	6	0
grnti7	green ti	7 gal	06/09/96	140.00	234.00	8	5	6	0
ficus14	ficus green gem	14 in	06/12/96	90.00	150.00	7	7	20	2
ficus17	ficus green gen	17 in	06/12/96	240.00	400.00	6	4	12	0
margi	marginata	2 gal	06/11/96	45.00	75.00	16	13	12	0

POTTERY INVENTORY

Code	Desc	Size	Features	AveNet$	List$	QIS	IS/OB
TCP50	t/c pot	50 gal	hand painted	53.95	110.95	6	3
GP50	granite pot	50 gal	oblate	110.00	195.00	5	3
TCF03	t/c figurine-swan		glazed	25.50	45.00	3	1
MBB30	marble bird bath	30 in	Doric base	128.50	250.00	2	1
GFN48	granite fountain	48 in	fluted basin	457.50	750.00	1	1

MATERIALS INVENTORY

Code	Desc	Unit	AveNet$	List$	QIS
CBRK5	decorative cedar bark	bag (5 cu ft)	7.50	15.95	53
CRGRN	crushed granite	yard	7.50	14.00	12
PGRV	pea gravel	yard	8.00	20.00	7
GRV1	1" gravel	yard	5.90	12.00	18
TSOIL	top soil	yard	12.50	20.00	10
PBLKG	patio block-gray	each	.56	.84	94
PBLKR	patio block-red	each	.56	.84	123

Figure PAS-7: Excerpt from *NBD* Inventory Report

"Okay," Tamara continued. "I prepare a Design Budget after a preliminary meeting with a client. Again, experience is the main factor in my ability to estimate the number of hours required to prepare the sketches and the bid. We designers are allowed to invest no more than 10 percent of the estimated bid on design activities. So the first thing I have to do is to estimate the bid amount based on my preliminary meeting with the client. Then I calculate 10 percent of that and divide that figure by 40 -- my cost per hour -- to come up with an estimate of the maximum number of hours I can spend on the bid process."

Labor Type	Labor Price/Hour	Labor Cost/Hour
Production worker	30.00	18.00
Designer	65.00	40.00
Equipment operator (contracted labor)	65.00	45.00
Botanist	75.00	50.00

Figure PAS-8: Excerpt from Labor Report

"Whoa," one of the students interjected. "Could you give us an example of that?"

"Sure. You have copies of the forms used for the London Square Mall project, right? Okay, in that case, the bid estimate was $7500. Ten percent of that is $750. Divide 750 by 40, giving about 18-19, the maximum bid process hours."

"Thanks. That's much clearer."

"Our daily work report is pretty straightforward. All I do is indicate the number of hours I've spent each day on each project. I give the Design Budget to Cheryl for approval and the Designer Daily Work Report to Connie."

"Okay. If you can't think of any other details about how the designers estimate and report costs, do you have any complaints about the current system? Are there any improvements you'd like to suggest?" Connie asked.

"Not that I can think of now. But I'll let you know if I come up with any other ideas."

Interview with Bill Johnson, Greenhousing Clerk

(see Bill's interview in "An Inventory Management System for Natural By Design," pages 34 - 35. Note that you can ignore the discussion of the shipping list as you study the Project Administration System.)

Interview with Monica Goce, Lead Production Worker

(see Monica's interview in "An Inventory Management System for Natural By Design," pages 35 - 39.)

Interview with Sue Kaufman

Connie peeked into Sue's cubicle. "Good morning, Sue. Are you ready for us?"

"Sure," Sue replied. "Just let me close this worksheet. Why don't you bring some chairs over for your friends while I finish up."

"Okay, we're ready if you are," Connie said, after everyone was seated. "We want to start today by discussing how you convert a bid into a Production Plan (see Figure PAS-2)."

"Sure. It's not a particularly difficult task. Just time consuming and tedious. But I'm happy to say that it's a little less so now that most of the designers are using your bid template (see desbid.xls). I have the designers give me a hard copy and a soft copy of each bid. That way, I'm able to copy shared data from the bid spreadsheet into my production plan spreadsheet (see prodplan.xls). You know, data such as project information, item code and quantity, etc. Keri gives me a copy of each bid on which she has noted the project team and project code (e.g., LS Mall). Then I refer to the latest inventory report (see Figure PAS-7) to determine the cost of each material requirement and enter it in the appropriate cell. That pretty much takes care of the material requirements portion of the plan; the spreadsheet does all the calculations. That is, assuming that you're not interested in delivery and installation scheduling."

"That's right," one of the students responded. "For now, we want to focus just on cost tracking. Stan and Keri suggested that including the scheduling procedures would make our investigation too complex. So we need to consider only the cost data and the procedures for tracking it."

"I have a question about the material requirements portion," another student interjected. "We've already interviewed Monica Goce. She mentioned that sometimes the quantity of an item actually used in a project is changed without revising the production plan or bid. In your experience, is the quantity of an item always exactly the same on the Production Plan as it is on the Design Bid?"

"Good question. Sometimes I need to modify the quantity estimates. For example, a designer might estimate that a project requires a yard of top soil, but looking at the blueprint, I see that a yard and a half is needed. I'm authorized to make such changes. We're only talking about a $6 change; revising the bid and plan and going through the approval process would be a waste of time and money in such a situation."

"Are there any other times that you change the quantities or types of items?"

"Sure. The designer may have recommended 1-inch gravel, but, after consultation with me and/or with the client, decides that pea gravel would be more appropriate. Again, we just go ahead and make the change. Or we may have trouble getting good quality specimens of a particular plant and decide to substitute a different plant. If it's a significant change in terms of dollars or appearance, we get approval from Cheryl, Keri and the client."

"Okay then, the quantity and type of item is not always exactly the same as on the approved bid."

"Right. The same is true of labor requirements. I don't usually have reason to change the design labor estimates, but, especially with a new designer, I may need to change the production labor estimates."

"Why isn't every bid sent to you for approval first, given that you so often have to change the estimates?"

"Oh, no, I don't change the estimates very often. I didn't mean to give you that impression. It's very rare that I have to make a change, and even more rare that the change I make is significant. In fact, most new designers work with me as they prepare their bids, so the kinds of significant or frequent changes you're envisioning just don't arise often enough to warrant my perusing every bid before it's approved."

"Okay. Thanks for clearing that up. You were about to tell us how you prepare the labor requirements section."

"Yes. That's a more involved process because I need to assign workers to each task and estimate the number of hours required. Again, making these estimates has become easier with experience, but, in some ways, it's still a seat-of-the-pants guess. For example, installing a large plant such as a laccospadix australasica palm generally requires two workers and takes about one to one and a half hours per plant. In the sample production plan you were given, you'll notice that two workers, Monica and Bert, were scheduled for eight hours -- four hours each -- to install three lacco palms."

"That's interesting. But I think we've become sidetracked into <u>how</u> you estimate hours. Since our main concern is tracking costs, we need to know how you derive your cost data. For example, how do you determine the cost per hour of a production worker or a designer?"

"Stan has defined several types of employees and has computed the average total hourly cost of each employee type (see Figure PAS-8). For example, including hourly wage, benefits, and other costs of employment, a production worker costs *NBD* about $18 per hour; a designer costs about $40 per hour. So, from your perspective, all I do is multiply the number of hours to be worked by each employee type by the hourly cost of that employee type to derive the estimated production labor costs. But I beg to differ with this simplification. I know that at the higher administrative levels, they think in terms of projects. Well, I do too. But I also think in terms of individual workers, and I need some way of tracking whether my workers are performing their jobs efficiently and effectively. I want to know if a worker consistently takes 25 percent more time to complete a task than what I estimated. I want to know if a worker's work has to be redone."

"How do you get this information?"

"For the most part, I don't. But if I notice that projects that a particular production team works on tend to go over budget, I'll pull out the PDW reports for those projects and manually compare estimated hours to actual hours for each task. It's a tedious, mind-numbing process, but one I feel compelled to perform on occasion."

"Okay, I think we can address that problem. Assuming all the PDW data is stored in a database, it would be pretty easy to generate a report sorted by project or by worker and comparing estimated hours to actual hours."

"You could do that?" Sue asked with a note of suspicion in her voice. "You're not just promising me the moon, are you? A friend of mine at another company said that computer consultants always promise more than they can deliver."

"Well, we're talking about a finer level of detail and a report that we'd have to have approved by management. But, if they give the go-ahead, it's certainly feasible. In fact, we could generate an exception report listing only those instances where the estimated-to-actual-hours ratio deviated by a certain percentage. And, now that I think about it, we could even generate a report showing the average hours required for a particular task for a particular size project."

"Wonderful! If you could do that, I'd lobby very hard to have it included in the project. I've often wished that I had some way of verifying the accuracy of my labor estimates. If I call a worker to task for what I consider unacceptable productivity, I'd like to have some hard numbers to back me up."

"Okay, we'll see if Stan and Keri will include that in the system boundary. But, for now, we need to focus our attention on our given agenda, which is facilitating and improving the accuracy of cost tracking. You've said that completing the production plan is a tedious, time-consuming process. Is there anything you can think of that would make the process less burdensome?"

"I suppose the usual things would help. You know, things like, I enter the code and quantity and the system fills in the description, unit cost, and extended cost. Also, anything that allows me to reuse data already existing about the project -- like from the bid. But that's about it. Connie's spreadsheet was certainly a step in the right direction; it takes care of a lot of the accuracy problems. But what you've said makes me very interested in being able to generate a variety of reports. I'm not sure you could do that with Connie's spreadsheet."

"Probably not. We'll talk to Stan and Keri and see what we can do."

Interview with Cheryl Poy, Keri Yamaguchi, and Stan Felton

"Sorry we're late. Our group meeting ran long," Connie apologized as she and her team members were seated in Stan's office.

"No problem, we all just got here ourselves," Keri said. "I think it's wonderful what your team is doing for *NBD*. I hope it's a good learning experience as well."

"It is," Connie said. "It's really amazing how everyone seems to have a different view of things, a different agenda. I've worked here for almost four years and always thought I knew the project administration procedures inside and out. But I'm learning a lot."

"Sorry to curtail the chitchat, but I've got an off-site meeting in an hour. So I think we should get right to the topic at hand," Cheryl interjected.

"Sure thing," Connie responded as she took out her project notebook. "The main topics we need to cover with you all are how you use the Project Administration Report (see Figure PAS-6) and what other reports you need to make it easier to track project costs. Let's begin with the Project Administration Report. How do each of you use this report? Stan, let's start with you."

"Let me tell you that the accuracy of our bid and production estimates can make or break *NBD*. If our designers consistently under bid or Sue underestimates production costs, we're in serious trouble. So my primary thoughts as I examine the PA report are 'How are we doing? How accurate are our estimates? Are we staying within the 50 to 70 percent range for production costs to bid amount? Are designers investing a reasonable amount of time in the bid process?' So, for example, when I looked at the report for June 18, I was very upset to see that design costs for the Fremont project were over 300 percent higher than estimated. Then I realized that there was a typo -- 100 should have been 1000 -- and breathed a sigh of relief! Not to cut on you, Connie, but I sometimes worry about the accuracy of this report. How many 'typos' or logic errors are made on the forms used to generate the report and on the report itself? How confident can I be in the information presented to me?"

Connie shook her head. "Reminds me of what my first MIS instructor always used to say: 'Garbage in, garbage out.' I understand your concern; that's why I've encouraged the designers to use the design bid spreadsheet and Sue to use the production plan spreadsheet to improve the accuracy of calculations. I try to verify all the figures before I use them to produce reports, but sometimes I just simply don't have time to double-check all the calculations. Anyway, if I understand you correctly, Stan, you use the report as a control device and to get a general sense of the profitability of the company."

"In a nutshell, yes."

"How about you, Keri?"

"I also use the report to gauge the company's profitability, and I'm also concerned about the accuracy of estimates. But I primarily look for trends over time. I'll often pull out all the reports for several weeks to see if any designers are consistently over- or under-estimating bid amounts. Obviously I'm more concerned about underestimates, but overestimates can be a problem too. We don't want to overbid if doing so ends up costing us business."

"The underestimates are my primary concern," Cheryl interrupted. "I especially pay attention to projects designed by new designers or unusual projects that differ substantially from anything *NBD* has done before. So I've been tracking the Fremont project closely; it's a fairly large project being designed by one of our recent hires, and it's a new kind of design for *NBD*. We haven't done a formal garden for anyone in the time that I've been here."

"As I'm listening to you," one of the students said, " I can't help but think that, in some ways, this report provides the wrong information, or, at least, not all the information you need, organized in the way you need. Ms. Yamaguchi, you noted that you sometimes pull out several reports to spot trends. Ms. Poy, you need to be able to determine how effectively each designer is estimating bids. This report doesn't really give you this information."

"No, it doesn't," Keri responded, "but I didn't want to ask Connie to prepare two or three reports from the same data. She's already working overtime as it is."

"It seems to me that you need some kind of database system," another student suggested. "Connie has said that she planned to develop a spreadsheet to generate this report, but after completing a database course, I've learned that a spreadsheet doesn't provide the reporting flexibility that most businesses need. The report itself might contain fewer calculation errors, but Connie would have to create a new spreadsheet and re-enter the data for each report. With a database, you could store the data in one format and generate a variety of reports from the same data. Ms. Yamaguchi could get a report showing trends over time or trends for particular types of projects, Ms. Poy could get a report sorted by designer, and Mr. Fenton could get a report showing the overage/underage for each project or for all projects for a given time period. In fact, you could generate an exception report listing projects that missed the mark by a certain percentage on a certain variable, for example, the implementation costs to bid amount ratio."

"That would be great," Cheryl commented, "but can you develop such a system?"

"Probably not all of it, but we could identify the requirements and get enough of it done so that you can evaluate whether it's worth hiring a consultant to finish. In other words, we'd develop a prototype that shows the general design of the database and the data entry screen and reports. Then you'd evaluate the prototype, suggest modifications, estimate the costs of developing the complete system, and, if you decided the benefits justified the costs, hire someone else to do the full implementation."

"That sounds reasonable," Stan concluded. "The questions and issues your team has raised during the analysis have been very useful, making me rethink some of the things we do. So, even if we decided not to implement the full system, the opportunity to re-evaluate our business processes more than justifies the time we've invested in the project. What do you say, ladies? Keri?"

"I fully agree. We've got nothing to lose. Cheryl?"

"Absolutely. I think it's about time *NBD* enter the information age. Of course, I have a hidden agenda. Once we have the computer resources in place, I'll be lobbying to get a CAD system for our designers. Drawing sketches and blueprints by hand is too time consuming."

"Don't jump the gun, Cheryl," Stan cautioned with a smile. "Let's learn to walk before we try to run the hurdles."

Interview with Connie Nguyen

"I handle reams of paper every week. To achieve some degree of order and easy access to documents, I maintain an expandable file folder for each project. In each project folder, I maintain all versions of the design bid and production plan, filed sequentially by date. I also include the design budget and any revisions to it, and every PDW report and DDW (designer daily work) report, again filed sequentially by date. As these forms come in -- and as time allows, I double-check the calculations and note any corrections on the form. I usually set aside at least two hours every day for compiling the cost data for projects. To help me compile the data, I created a spreadsheet template to track the costs of each project each day (see Figure PAS-9).

"To prepare the Project Administration Report (see Figure PAS-6), I compute hour and cost totals for each cost type for each project for that week -- a week runs Monday through Friday. Then I add these to the appropriate actual hour and cost totals from the previous week's report. If the project is still in the design stage, I subtract the grand total from the hours and dollars remaining figures from the previous week's report. Once I've finished all my calculations, I call up the Project Administration document template in Word and fill in all the details for the report."

Project:	LS Mall	
Bid Amount:	$7457.50	
	Estimated	Actual (To-Date)
Production Materials Costs:	$3240.00	$3255
Bid Hours/Cost:	12/$480	12/$480
Production Design Hours:	10/$400	10/$400
Production Labor Hours:	40/$990	41/$1008
Total Hours/Cost:	62/$5,110.00	63/$5,143

	4/15	4/21	4/25-5/2	5/23	6/14	6/15	6/16	6/17	6/18	Total
Materials						800.00	1350.00	1105.00		3255.00
Design Hours-Bid	2	5	5							12
Bid Cost	80.00	200.00	200.00							480.00
Design Hours-Prod				5		3	1		1	10
Prod. Design Cost				200.00		120.00	40.00		40.00	400.00
Labor Hours					6	10	9	16		41
Labor Cost					270.00	288.00	162.00	288.00		1,008.00

Figure PAS-9: Sample of Connie's Data Compilation Form

A SALES AND ACQUISITIONS PROCESSING SYSTEM FOR BOOKPLACE, INC.

I. Background

A. Organizational Overview

BookPlace, Inc. is a wholly owned subsidiary of a major U.S. conglomerate that also operates companies in publishing, consumer products, and entertainment. BookPlace is organized as a divisional bureaucracy; that is, it is divided into three divisions, each with its own purchasing, sales, accounting, and other functional areas and each reporting to a central executive level responsible for administration, human resource management, long-term planning, legal affairs, and finance. The three divisions of BookPlace are Popular Books, Textbooks, and General Books. The General Books division is the "stepchild" of the BookPlace organization. Employees of this division commonly joke that "General" is an euphemism for "none of the above." In other words, if a book is not a popular book -- i.e., the kind sold at B. Dalton and other bookstores -- or a textbook, it's a "general" book. As the stepchild division, General Books has been largely ignored and underfunded by BookPlace management.

The General Books (GB) division employs twelve people, including four managers: Ambrose Yurkel, the divisional head who allocates resources and oversees all activities in the GB division; Jonathan Wassenaar, who manages Sales and Acquisitions; Christina Domingo, the head accountant who maintains the division's Accounting records; and Farah Faisal, the Communications director responsible for all divisional communications. Jonathan, Christina, and Farah report to Ambrose, who, in turn, reports to the CEO of BookPlace, Theodore MacCormick. The other GB staff include Yvette Williams, Mr. Yurkel's executive assistant; three Sales and Acquisitions clerks -- Tran Pham, Lisa Edmonds, and Jaime Guzman -- who are at Jonathan's beck and call; an Accounting clerk, Greg Loi, who reports to Ms. Domingo; and three Communications clerks -- Bobbi Jo Murphy, Katrina Rhu, and Sam Rao.

B. History of the Organization

Before its buyout by the conglomerate four years ago, BookPlace was largely a Yurkel family enterprise. Ambrose's grandfather, Irvin, established Yurkel Books in Brooklyn, New York, in 1916 as a family business that sold books, magazines, and newspapers and catered to the needs of a largely Eastern European immigrant neighborhood. In 1945, Ambrose's father, Samuel, returned from the war in Europe to take over the family business. Samuel was a prodigious businessman who grew Yurkel Books into the BookPlace chain with twelve stores in the New York City metropolitan area. Both Irvin and Samuel were bibliophiles -- collectors of rare books -- who maintained the company's rare book acquisitions and sales operations more for love than for profit. Distracted with efforts to expand the business, Samuel hired Jonathan Wassenaar to manage rare book operations in 1963.

Failing health led Samuel to turn the operation over to his son Ambrose in 1981, though Samuel remained as the titular head of the business. Ambrose, trying to follow in his father's footsteps, leveraged the company to open stores throughout the Northeast. But a lingering recession and its accompanying decline in sales put BookPlace on the brink of bankruptcy, unable to repay the loans used to fund the expansion. That's when the conglomerate stepped in. Samuel agreed to sell the business with one stipulation: his son, Ambrose, would continue to manage the company until Ambrose chose to leave or to retire.

The conglomerate kept its contractual promise -- sort of. After the buyout, Ambrose was demoted to head of the General Books division, and Ted MacCormick was hired to run BookPlace, Inc. Needless to say, Ted and Ambrose have few kind words to say about each other; this rancor no doubt explains GB's stepchild status. For example, while the other divisions have been moved into a new office building, reorganized, modernized, and provided extensive funding to employ information technology to improve efficiency, General Books continues to languish in the four-room office suite where Irvin Yurkel used to enjoy a cigar at the end of each day.

II. Current GB Systems

Currently, GB has just four 386-PC clones equipped with 120 MB hard drives and one daisy-wheel printer that clatters noisily along until end-of-quarter reports are being processed, at which point it suffers a major breakdown, requiring much 'handholding' (i.e., service) and causing much handwringing among the GB staff. Yvette and Christina each have a PC, and the Communications area has two. All four PCs are linked to the printer through a 4-position (A-B-C-D) switch box. Thus, if Yvette wants to print a document, she must walk over to the switch box and set the switch to A; if Christina wants to print a worksheet, she must walk over to the switch box and set the switch to B; etc. An early version of Word for Windows is installed on all PCs; in addition, Christina's PC has an early version of Excel. Because of its limited technology, most of the division's information processing is performed manually.

A. Current Sales and Acquisitions Systems

Jonathan Wassenaar is known throughout the world for his ability to locate and sell rare manuscripts. His success, in large measure, results from his prodigious memory, his fluency in seven languages, and his relationships with numerous rare book dealers and collectors around the world. In fact, Jonathan's knowledge is so vast that the world's museums and libraries consult him to locate rare items. As manager of Sales and Acquisitions, Jonathan oversees several activities: maintaining records of clients, client contacts, sources, source books of special interest, and current book inventory; processing acquisitions and sales; and generating client correspondence, the monthly newsletter, and management reports.

Client Tracking. The key to Jonathan Wassenaar's success is his extensive client base. Whenever he receives an order or inquiry from a new client, Jonathan records the client's name, address, phone number, and interest on a 3x5 note card and gives it to one of his clerks. The clerk assigns the new client a client ID number, prepares a folder marked with the customer's name, and files it alphabetically in the client file; the clerk also adds the client ID to the 3x5 card and files the card alphabetically on Jonathan's client rolodex. (See Figure BP-1 for an excerpt of Jonathan's client list.) Currently, Jonathan's client file includes almost 2,500 customers.

As established clients contact GB or make purchases, Jonathan or one of his clerks notes each customer purchase and inquiry in the appropriate client folder; they also keep track of any other contacts with each customer, such as telephone conversations conducted and letters sent. A sample of Jonathan's client record notes is shown in Figure BP-2.

At the end of each quarter, the Sales and Acquisitions clerks review the client folders to identify customers who made a purchase during the previous three months. Then Jonathan composes customized letters to these customers. Typically, he writes three or four different types of letter. When writing to particularly valuable clients, he personalizes a letter by referring, by name, to a recent purchase. For example, he might start a letter with the sentence "I hope you're enjoying the *Huckleberry Finn* First Edition you acquired last month," and then continue to mention a related book currently available. These hand-written letters are given to the communications clerks, who word process the letters and send them to the clients. To maintain contact with all his customers, Jonathan also prepares a monthly newsletter, which, again, is processed and mailed to every customer each month, by the communications staff.

Client ID	Name	Address	Phone	Interest
BA089	Marilyn Baker	31 James Woods CT, Manchester England	011-0522-75 88 99	Child Lit
CH034	Frank Charles	31762 N. Sunset, Los Angeles CA 96666	(213) 757-8800	SciFi
DO057	Jonathan Doyle	2943 Cordell Ave, Cleveland OH 44113	(216) 779-5050	19th Cent Am.
KI029	Kenneth King	485 High St., Victoria 3181 Australia	061-0347-88 42 17	20th Cent Am.
LA083	Steve Lawrence	Castle St., Salisbury, Wiltshire, England	011-7033-90 75 22	Child Lit
PH021	Stone Phillips	65 George St., Edinburgh EH2 2JL UK	044-0311-87 09 66	Modern
ST103	James Stargardt	Rade-Strasse 10, Hamburg Germany	049-4033-69 90 22	19th Cent Am.

Figure BP-1: Excerpt of the Client List

Marilyn Baker (BA089)

Contacts

Date	Type	Content
04/01/96	Letter	items on ChildLit list
04/22/96	Phone	Peter Rabbit available
04/27/96	Phone	Add 'Jemima' to collection?

Sales

Date	Invoice#	Author	Title	Sales Price
04/25/96	01173	Beatrice Potter	The Tale of Peter Rabbit	$50,000 + $7,500
04/27/96	01185	Beatrice Potter	Jemima Puddle-Duck	$765

Frank Charles (CH034)

Contacts

Date	Type	Content
04/01/96	Letter	Items on SciFi list
04/25/96	Phone	'Stranger' available; interested?

Sales

Date	Invoice#	Author	Title	Sales Price
04/28/96	01198	Heinlein	Stranger in a Strange Land	$700 + $105
		Heinlein	The Green Hills of Earth	$120

Figure BP-2: Samples of Jonathan's Client Record Notes

Sales and Acquisitions Processing. As noted above, Jonathan has just under 2,500 clients. Last year, these clients submitted about 7,000 orders involving 15,000 items from about 200 sources. Currently, Jonathan devotes a 4-drawer file cabinet to his annual sales order records (a sample invoice is shown in Figure BP-3), which are filed sequentially by sales date in folders marked with the month and year. Out-of-date folders (i.e., records of orders processed more than 12 months earlier) are moved into a storage area in the basement of the building.

The sales order processing activities are initiated when a client submits an Intent to Purchase: a letter or purchase order that gives the client's billing and shipping information and that identifies the items to be purchased. The Intent to Purchase constitutes a legal offer to buy and forms the sales contract. To process the sale, one of the Sales and Acquisitions clerks completes an invoice form in triplicate; each invoice form is preprinted with an invoice number. First, the clerk enters the current date on the invoice form. Then, the clerk compares the client's address on the Intent to Purchase to the client's record in the Client file folder and updates the client address, if necessary. The clerk then enters the client's current address on the invoice form; the clerk also enters the client's account number or credit card number, if appropriate. If the client submitted a purchase order instead of an Intent to Purchase, the clerk enters the client's PO number and data for reference.

Next, the clerk accesses the appropriate book records from the book files to enter data about each book being sold -- including book code, author, title, publication date, edition, and selling price. Then, the clerk calculates the invoice subtotal, sales tax, shipping and handling charges (typically, $7.50 per item; more if the books are sent insured express delivery), and total and enters these figures on the invoice form. The clerk photocopies the invoice form and places the photocopy in the Sales Pending tray. All three copies of the invoice form and the Intent to Purchase are then forwarded to Accounting.

If the client must be billed for the order, Accounting sends the white and pink copies of the invoice to the client and files the yellow copy and Intent to Purchase in the Accounts Receivable folder. When the client submits payment along with the pink invoice copy, Accounting retrieves the yellow copy and Intent to Purchase from its Accounts Receivable file and marks both copies "Paid." Accounting files the yellow paid invoice copy in its Payments folder and forwards the pink copy to Sales and Acquisitions. A Sales and Acquisitions clerk accesses the appropriate book records, annotating the records to include the sales date and the name of the client who purchased the book. The annotated notecard is then filed alphabetically by author in the Books Sold file. Finally, the clerk ships the pink invoice copy and book(s) to the client, retrieves the photocopied invoice from the Sales Pending tray, and files it chronologically in the appropriate Sales Order folder.

01198

BookPlace, Inc.
2955 E. First Street, Suite 104
New York, NY
(718) 444-1212

Client Name:	Frank Charles	**Order Date:**	04/28/96
Client Address:	31762 N. Sunset	**Order Number:**	**01198**
	Los Angeles, CA 96666	**Account Number:**	MC408197853421
PO Ref. Date:		**PO Ref. Number:**	

Line Items

Book ID	Author	Title	Pub Date/Ed.	Sales Price
HEI10321	Heinlein	Stranger in a Strange Land	1961/1st	$805.00
HEI00980	Heinlein	The Green Hills of Earth	1951/1st	$120.00

Subtotal	$925.00
Sales Tax	7.40
S/H Charge	15.00
Total	$947.40

Figure BP-3: A Sample Sales Record (Invoice)

Jonathan also purchases items with the intent to resell them. The GB division sets aside a budget each year for purchasing outstanding values that Jonathan identifies as he studies the rare book dealer catalogs. To process an acquisition, Jonathan completes a purchase order form (see Figure BP-4) in triplicate and forwards it to Ambrose for approval. Once approved, Ambrose gives the form to Accounting, which sends the white form to the seller along with a check for the purchase amount. The yellow form is filed in accounting, and the pink form is returned to Sales and Acquisitions, where it is filed in the current inventory folder.

Source and Item Tracking. Jonathan and his clerks also maintain records of his sources and the books he has available for purchase. Jonathan maintains a folder for each of his sources. These sources include auction houses and individual book dealers who trade in rare books and manuscripts. Jonathan maintains data about each source's name, address, phone number, and contact person, as shown in Figure BP-5. Each time Jonathan purchases an item from a source or acts as a broker between a source and one of his clients, he or one of his clerks not only files a copy of the purchase order, but also records the transaction in the appropriate Source and Client files, although sometimes they forget. A sample of these source notes is shown in Figure BP-6.

09432

BookPlace, Inc.
2955 E. First Street, Suite 104
New York, NY
(718) 444-1212

Source Name:	Swain Galleries	**Order Date:**	04/21/96
Source Address:	104 East 45th St.	**Order Number:**	09432
	New York, NY 10010	**Account Number:**	BP9624

Line Items

Item Number	Author	Title	Pub Date/Ed.	Purchase Price
HEM00128	Hemingway	A Farewell to Arms	1929/Scribner's	$125.00
KER00876	Kerourac	On the Road	1957/1st	$35.00

Subtotal	$160.00
Sales Tax	12.80
S/H Charge	15.00
Total	$187.80

Figure BP-4: A Sample Purchase Order

Source Code	Source Name	Address	Contact
bbc	Ballentine Book Company	4233 S. Charles, Baltimore MD 21218	Bob Stewart
Ck	Christal S. Kensington	8543 Broton Rd, London SW1P 6QT England	Diane Powell
CNY	Christian, Munson & Wells	50223 5th Ave, New York NY 10022	Peter Struthers
K	Kanelle & Schwinn	4830 Bielefeld 1 Frankfurt 9 Germany	Arne Kanelle
P	Sachs Parke Bernet, Inc	31334 Yonkers Ave, New York NY 10021	Sara Milne
pba	Pacifist Book Auction	1924 Levi St, San Francisco CA 94133	Steve Michel
pn	Philberts, Son & Nease	101 James Bond St, London W1Y England	Austin Nease
S	Sackesby Italia	Via Pietro Mascagni, 20122 Milan, Italy	Gina Berti
sg	Swain Galleries	104 East 45th St, New York NY 10010	Cal Swain
wa	Waverlee Book Auctions	4333 Cornell Ave, Baltimore MD 10814	Betty Waverlee

Figure BP-5: An Excerpt of Jonathan's Source List

SWAIN GALLERIES

Acquisitions

Date	Book ID	Author	Title	Purchase Price	PO #
04/21/96	HEM80143	Hemingway	A Farewell to Arms	$125.00	09432
04/21/96	KER00876	Kerouac	On the Road	$35.00	09432
01/26/94	HEI00980	Heinlein	The Green Hills of Earth	$50.00	08901

Brokered Sales

Date	Book ID	Author	Title	BP PO#	Invoice#	Sales Price	Fee
04/25/96	POT12075	Beatrice Potter	The Tale of Peter Rabbit	09433	01173	$57,500.00	$7,500.00

KANELLE & SCHWINN

Acquisitions

Date	Book ID	Author	Title	Purchase Price	PO #
12/07/93	HEM12002	Hemingway	The Old Man and the Sea	$15.00	08879

Brokered Sales

Date	Book ID	Author	Title	BP PO#	Invoice#	Sales Price	Fee
04/28/96	HEI10321	Heinlein	Stranger in a Strange Land	09435	01198	$805.00	$105.00

PHILBERTS, SON & NEASE

Acquisitions

Date	Book ID	Author	Title	Purchase Price	PO #
02/06/94	POR99923	K.A. Porter	Pale Horse, Pale Rider	$80.00	08898
02/22/94	POT00222	Beatrice Potter	Jemima Puddle-Duck	$650.00	08910

Sales

Date	Book ID	Author	Title	Sales Price	Invoice #
04/26/95	JAC33333	Jackson	The Haunting of Hill House	$100.00	01181

Figure BP-6: Samples of Jonathan's Source Notes

Jonathan reads the rare book dealer catalogs religiously to keep abreast of the market. When he finds a particularly good price on a book, he buys the item on speculation, assuming that he'll be able to sell it to one of his clients. Other times, he just notes that particular items are available from particular sources. Jonathan maintains 3-by-5 notecards with information about these items in card catalog files salvaged from a closed city library. In addition to the book's identifying number, author and title, each card lists the place and date of publication, the edition, the type of binding, a brief description of the book's condition (e.g., "original dust jacket," "water damage along top of first several pages," and so on), any distinguishing features (e.g., "signed by author," "inscribed to President Truman," and so on), and the book's source and current price. Figure BP-7 shows examples of books in Jonathan's collection or on his "good buys" list.

Sample Reports. Jonathan uses several reports to keep track of his clients, sources, orders, and books. Most of these reports are manually compiled and typed by his clerks. Figure BP-8 shows an excerpt from the Client Contact Report, which summarizes Jonathan's communications with his top three hundred customers over the prior three months. Figure BP-9 shows an excerpt from a Sales Report, which lists all the sales for a particular period by type and gives a total sales figure for that period. Figure BP-10 shows an excerpt of the Book Inventory Report, which lists information about the books that Jonathan owns and gives the total value of his current inventory.

Generating these reports from Jonathan's manual recordkeeping system is a monumental task. Retrieving and refiling folders and cards keeps one of Jonathan's assistants busy full time! Furthermore, because his current manual system includes so much redundant data, the data used to generate these reports are not always consistent, accurate, and up to date.

B. Other GB Systems

The Sales and Acquisitions manual systems involve a major portion of GB's workload. However, Yvette, Christina, and Farah also operate partly-automated systems in their areas. Yvette uses her microcomputer primarily to process Ambrose's correspondence. All letters, intra-firm memos, and other documents are saved in separate directories on Yvette's hard drive. Yvette also maintains Ambrose's schedule on a desk calendar; during particularly busy times, she types his daily schedule using Word and then prints it and places it on his desk first thing in the morning.

Book ID	Author	Title	Pub Place	Pub Date	Edition	Binding	Condition	(Original) Source	Current Price
Books Owned by BookPlace									
AND10754	Anderson	Winesburg, Ohio	NY	1919	1st, 1st iss.		hinge crck'd	P	$80
FAU20543	Faulkner	Absalom, Absalom!	NY	1936	1st	original	Epstein copy	Ck	$1,800
FAU39485	Faulkner	Light in August	NY	1932	1st	cloth	chipped	pba	$120
HEI00980	Heinlein	The Green Hills of Earth	Chicago	1951	1st		d/j, edge wear	sg	$120
HEM00128	Hemingway	A Farewell to Arms	NY	1929	Scribner's	orig wraps	worn	sg	$300
HEM12002	Hemingway	The Old Man and the Sea	London	1952	1st English		d/j	K	$50
JAC33333	Jackson	Haunting of Hill House	NY	1959	1st		d/j	bbc	$100
KER00876	Kerouac	On the Road	NY	1957	1st		d/j	sg	$80
MEL00988	Melville	Moby Dick	London	1851	1st	mor gilt	spine worn	CNY	$17,000
POR99923	K.A. Porter	Pale Horse, Pale Rider	NY	1939	1st		d/j	pn	$110
POT00222	B. Potter	Jemima Puddle-Duck	London	1908	1st	16mo	orig binding	pn	$765
Books Available through other Dealers									
FAU21005	Faulkner	Absalom, Absalom!	NY	1936	1st		rubbed d/j	bbc	$300
FAU40109	Faulkner	As I Lay Dying	NY	1930	1st	orig cloth	d/j	wa	$275
HAW01269	Hawthorne	The Scarlet Letter	NY	1850	1st	orig cloth	Inscr	P	$21,000
HEI10321	Heinlein	Stranger in Strange Land	NY	1961	1st		d/j	K	$700
HEM80143	Hemingway	A Farewell to Arms	NY	1929	1st	orig cloth	covers spotted	bbc	$1,600
HEM09231	Hemingway	For Whom the Bell Tolls	NY	1940	1st	orig cloth	Epstein copy	sg	$4,000
HEM34087	Hemingway	For Whom the Bell Tolls	NY	1940	1st	orig cloth	signed	bbc	$350
HEM20917	Hemingway	The Old Man and the Sea	NY	1952	1st in Bk		Inscr to WPalmer	sg	$3,400
HEM31956	Hemingway	The Sun Also Rises	NY	1926	1st, 1st iss.		d/j, Epstein copy	sg	$16,000
JAM20943	James	The Portrait of a Lady	London	1881	1st	orig cloth	rubbed, split	S	$1,700
JAM10358	James	Works	NY	1907-	NY Ed.	orig half cloth	worn, discolored	CNY	$3,800
KES42091	Kesey	Cuckoo's Nest	NY	1962	1st		d/j, Epstein copy	sg	$500
POT12075	B. Potter	The Tale of Peter Rabbit	London	1901	1st, 1st iss		1/250, Epstein	sg	$50,000
WHI23981	Whitman	Leaves of Grass	Brooklyn	1855	1st, 1st St.	orig cloth	Doheny-Manney	P	$29,000

Note: If BookPlace sells a book listed as available through another dealer, it generally adds a 15 percent commission to the listed price.

Figure BP-7: Sample Books on Jonathan's List
(see booklist.xls on project disk)

Client Contacts Report
1-Feb-96 to 30-Apr-96

DATE	TYPE	NAME1	NAME2	DESCRIPTION
3/1/96				
	L	Marilyn	Baker	items on ChildLit list
	L	Frank	Charles	items on SciFi list
	L	Jonathan	Doyle	items on 19C Am list
	L	Kenneth	King	items on 20C Am list
	L	Steve	Lawrence	items on ChildLit list
	L	Stone	Phillips	items on Modern list
	L	James	Stargardt	items on 19C Am list
3/5/96				
	P	Jonathan	Doyle	Rare 'Moby Dick' in stock
3/9/96				
	P	Kenneth	King	He called re 'Pale Horse'
		....		
4/22/96				
	P	Marilyn	Baker	'Peter Rabbit' available!
4/25/96				
	P	Frank	Charles	'Stranger' available
	P	Jonathan	Doyle	Competing bid 'Moby Dick'
4/27/96				
	P	Marilyn	Baker	Add 'Jemima' to collection?

Figure BP-8: An Excerpt from a Customer Contacts Report

Name	Invoice#	Author	Title	Price[1]	Revenue
			Sales Report		
			1-Apr-96 to 30-Apr-96		
Inventory Sales Directly to Clients					
Baker	**Marilyn**				
	01185	B. Potter	Jemima Puddle-Duck	$765.00	$765.00
Charles	**Frank**				
	01198	Heinlein	The Green Hills of Earth	$120.00	$120.00
King	**Kenneth**				
	01190	K.A. Porter	Pale Horse, Pale Rider	$110.00	$110.00
	....				
Inventory Sales through Dealers					
Philberts, Son & Nease					
	01181	Jackson	Haunting of Hill House	$100.00	$100.00
	....				
			Total Inventory Sales:	**$15,950.00**	
			Revenue from Inventory Sales:		**$15,950.00**
Brokered Sales					
Baker	**Marilyn (sg)**				
	01173	B. Potter	The Tale of Peter Rabbit	$57,500.00	$7,500.00
Charles	**Frank (K)**				
	01198	Heinlein	Stranger in a Strange Land	$805.00	$105.00
	...				
			Total Brokered Sales:	**$85,845.00**	
			Total Brokered Sales Revenue:		**$12,397.00**
			Grand Total Sales :	**$101,795.00**	
			Grand Total Revenue:		**$28,347.00**

Figure BP-9: An Excerpt of a Sales Report

[1] The Sales Price for brokered sales indicates the dealer's selling price. The price to the client is the Sales Price plus the finder's fee (listed as Revenue). For example, Marilyn Baker paid $57,500 for Peter Rabbit: $50,000 to cover the dealer's (sg - Swain Galleries) price plus $7,500 to cover BookPlace's 15 percent finder's fee.

Book Inventory Report
31-Dec-95

BOOKID	AUTHOR	TITLE	CURRENT PRICE	PURCH. PRICE	PURCH. DATE	PO#	ORIGINAL SOURCE
AND10754	Anderson	Winesburg, Ohio	$80.00	$50.00	10/23/93	08757	P
FAU20543	Faulkner	Absalom, Absalom!	$1,800.00	$1,000.00	04/19/92	07502	Ck
FAU39485	Faulkner	Light in August	$120.00	$70.00	02/01/94	08895	pba
HEI00980	Heinlein	The Green Hills of Earth	$120.00	$50.00	01/26/94	08901	sg
HEM12002	Hemingway	The Old Man and the Sea	$50.00	$15.00	12/07/93	08879	K
JAC33333	Jackson	Haunting of Hill House	$100.00	$45.00	02/06/94	08899	bbc
MEL00988	Melville	Moby Dick	$17,000.00	$10,250.00	05/12/87	05692	CNY
POR99923	K.A. Porter	Pale Horse, Pale Rider	$110.00	$80.00	02/06/94	08898	pn
POT00222	B. Potter	Jemima Puddle-Duck	$765.00	$650.00	02/22/94	08910	pn
		...					
		Total current value:	**$129,865.00**				

Figure BP-10: An Excerpt of a Book Inventory Report

Christina and Greg maintain spreadsheets of GB's accounts. These are stored on the hard disk and backed up each quarter to floppy. Christina also maintains the general ledger, generates payroll, and creates accounting reports (e.g., balance sheets, profit-and-loss statements, and so on), which are a combination of spreadsheets created using Excel and text created using Word.

Farah and the Communications staff are crippled by the availability of only two computer systems and one printer that they must share with other areas. Often times, they 'borrow' Yvette's or Christina's computer if it is not currently being used. The Communications activities are all word-processing oriented. They maintain separate directories for newsletters, client letters, and other documents on the hard drives. They try to keep the documents on the hard drives in sync by saving new document files to floppies and then copying them to both hard drives at the end of each day. In other words, to allow shared access to these files, current work documents are saved to floppies and backed up on the hard drives of both computers.

Sam Rao has recently begun copying current book inventory data from Christina's spreadsheets to disk and importing it into Word to facilitate preparation of the monthly newsletters. In addition, Bobbi Jo and Katrina have created a mail-merge client address file to facilitate generation of mailing labels.

III. GB Enters the Information Age

General Books seemed destined to languish in isolation, beleaguered by overflowing file cabinets and card catalog drawers. But then an industry award sparked a change in Ted MacCormick's attitude and launched GB into the Information Age.

A. Ted Has a 'Change of Heart'

At the annual convention of booksellers last week, BookPlace sponsored a table at the awards banquet. Normally, only high-ranking BookPlace executives attended the convention and its festivities. But, because the convention was being held in New York City and -- more importantly -- because it was rumored that the General Books division would receive an award, Ambrose Yurkel and Jonathan Wassenaar were allowed to attend as well. During the after-dinner awards presentation, the following scene occurred.

"Com' on, com' on -- Jeez, how many awards do they give out? Must be they're trying to get on everybody's good side, huh, Ambrose?" Ted MacCormick muttered, leaning back in his chair and sipping his champagne cocktail. "When's our turn? Or was this just a lot of hype to con us into buying a table?"

Ambrose didn't respond. What could he say. The man was a loot as far as Ambrose was concerned. Once a lout, always a lout. But he was also the boss, a very astute businessman who was not to be contradicted. So silence seemed the best tactic.

Another award presenter took the podium and began his award introduction. "Over the past eight decades, one company has been *the* source for information about rare books and manuscripts. Bibliophiles around the globe, when searching for an obscure title, knew that if it existed, Irvin Yurkel would find it. Later, it was Sam they turned to unearth lost works. Today they turn to Sam's son, Irvin's grandson, Ambrose Yurkel, and to his family's dear friend and associate, Jonathan Wassenaar. Today, Yurkel Books has become BookPlace, but I'm proud -- and, as a bibliophile myself, relieved -- to say that BookPlace has continued the Yurkel family tradition. Its executives and employees continue to perform an invaluable service in helping maintain and expand the collections of rare books housed in libraries and museums on all seven continents. Through their efforts, we can be assured that these priceless treasures are preserved for our children and our children's children ...."

"OK, already. Cut to the chase. Hey, Ambrose. What did you pay this guy not to mention my name in his treatise on the Yurkel tradition? I'm the one funding your 'humanitarian' pursuits, right? So how come I'm not at least commended for bankrolling your negative cashflow division?"

"…. It is with great pleasure that I ask Ambrose Yurkel and Jonathan Wassenaar to come forward to accept the 1995 Booksellers Service Award --"

"Let's go. I'm not on the marquee, but I'm going to put in an unscheduled appearance," Ted announced, standing and straightening his white tie as he herded Ambrose and Jonathan to the stage….

Early the next week, Ambrose and Jonathan were summoned to Ted's office to discuss plans to upgrade GB's information systems. As they rode the elevator to the BookPlace executive suite, Ambrose said, "Do you suppose that 'Mac the Knife' had some kind of religious conversion on the way to the podium last week? I can't believe he's actually going to put money into GB."

"'Mac the Knife?' Have you devised yet another name for our revered CEO?," Jonathan laughed. Then placing his hand on Ambrose's shoulder, he said, "Be honest, Ambrose. You just don't like the man. And it's not just that he has your job. You can't stand him because, not only does he have your job, but he's fifteen years younger, half a foot taller, fifty pounds lighter, and has more hair in his mustache than you have on your head."

As usual, Jonathan made Ambrose smile in spite of himself. Jonathan continued, "Anyway, to respond to your query, I doubt that this has anything to do with religion, unless you include the Church of the Almighty Dollar in your definition of that term. Ted's a sharp businessman and a marketing whiz. If you ask me, he sees this as a way to put BookPlace on the map, so to speak."

"Well, I guess we'll soon find out what has caused 'Tight Ted' to open his wallet," Ambrose sighed, as he and Jonathan stepped out of the elevator into the tastefully decorated foyer of the executive suite. A receptionist stepped forward and ushered them into Ted's office.

"May I offer you gentlemen some coffee or tea?" she asked, on her way out. "No? Okay, well, if I can get you something later, don't hesitate to ask."

Ted swept in through a side door. "Sit down, sit down. Ambrose. Jonny. It's good to see you again," Ted welcomed them. Still grasping Ambrose's hand, he continued, "Oh, Ambrose, don't make a face. I know that we don't see eye-to-eye. You'll never forgive me for being chosen to run 'your' company. But, let's face it, the award last week has given us a wonderful opportunity to put BookPlace on the map."

Hearing Ted repeat Jonathan's exact words, Ambrose and Jonathan shared a furtive look and then seated themselves in two leather chairs facing Ted's expansive desk.

Taking his position behind his desk, Ted plunged forward. "Anyway, I asked you gents up here to discuss a marketing plan I've been thinking about. You know that I've been thinking about expanding BookPlace, but, until now, we've lacked a defining feature to distinguish us from every other bookseller out there. Well, you guys handed me 'product differentiation,' or 'market niche,' or whatever right on a silver platter. 'BookPlace -- *The* Place for Books.' Not just the usual bestsellers and how-to-be-a-better-person sort of trash. But *any* book. 'If the libraries of the world come to us, why don't you?' You see where I'm headed with this? Jonny, you're a sharp guy. Do you see how this can benefit the company?

"Certainly, Mr. MacCormick. But I don't know whether it's appropriate to equate the casual book reader with the devout bibliophile. Why would Mrs. X come to BookPlace for the latest 'good read'? It's not as though she needs me to find books on the bestsellers list?"

"Yeah, I understand your point, Jonny, but you're confusing real benefit with perceived benefit. By offering an electronic book finding service, we differentiate ourselves from the B. Dalton's and Waldenbooks of the world. So what if most of our customers don't use the service. The thing is it's there and we can exploit it in a marketing sense. I have no illusions about making money from this service. Just breaking even would be an improvement over the status quo. Ambrose, don't give me that 'Literature for Literature's sake' holier-than-thou look. So I'm one of the great unwashed masses that can't tell a first edition Shakespeare from a John Grisham thriller. The thing is, I believe that we can sell more Grisham thrillers and more rare books if we implement this system. What do you think, Jonny?"

"I haven't your marketing expertise, Mr. MacCormick. But I can see the merits of your proposal, at least as it affects the GB division. I believe that we could at least break even if we didn't have to expend so much time and so many resources just maintaining client, source, and book records. Your idea of offering our booklist electronically -- and helping small, independent rare book dealers to do the same -- makes sense to me. I don't understand electronics, but I do understand that, in the business world, time is money. Currently, finding and promoting rare books takes too much time to make us any money."

"Great! So you're sold on the idea. Ambrose, how 'bout you?"

"All I know is that we can't go on the way things are. We're a nineteenth century outfit trying to survive in a twenty-first century world. As my son would say, 'Go for it ... ride the information wave.' Whatever that means. He's studying computers and business at NYU.... Anyway, I'm sure he'd think this was a great idea. I really don't know."

"Oh, Ambrose," Ted intoned, then rubbed his eyes as though trying to clear cobwebs. "OK, that's a 'yes' from Jonny and a 'maybe' from Ambrose. Good enough for me. Here's an outline of the plan. You'll notice that we've planned two project phases ...."

B. Outline of the Project Plan

The Sales and Acquisitions Processing (SAP) system project outline called for a three-phase implementation of information systems to support the General Books division. Phase One would focus on internal systems, redesigning and automating Sales and Acquisitions' processes.

Six months and $100,000 were allocated to this phase. Phase Two would upgrade GB's accounting and communication systems, including accounts receivable, accounts payable, general ledger, and so on. Three months and $45,000 were budgeted for this phase. Phase Three would exploit the Internet to create an electronic marketplace for rare books. This would involve setting up databases of 'Wanted to Buy' and 'Offered to Sell' records to be matched automatically by the system. If no matches were found, Jonathan would query records of his source's lists and pursue his usual tactics to try to find matches. This phase was allocated $100,000 and six months. In other words, within just over a year, BookPlace planned to transform the GB division and make it the marketing centerpiece for further corporate expansion.

A team of consultants was hired to analyze GB processes and to design and implement the Phase One system. The consulting team interviewed several end-users, the user-manager, and the user-sponsor. Transcripts of these interviews are provided here.

Interview with Ted MacCormick

Ted: OK, let's make this short and sweet. I've got a dozen other things to do this morning.

Consultants: Sure thing, Mr. MacCormick. We just need to check with you to verify our mission -- you know, the objectives, constraints, scope of the project --

Ted: Wasn't that explained in the document sent to your consulting group? Why do we have to discuss it again?

Consultants: Yes, it was explained in your request for proposal, but we need to take just a few minutes to verify our understanding. You know how two different people can read the same document and come to different conclusions. I promise that we'll be out of here in about 15 minutes. OK?

Ted: Shoot.

Consultants: OK. As we understand it, we are authorized to conduct the analysis, design, and implementation of Phase One of the project. Is that correct?

Ted: Sure is. If you do a good job, we'll probably keep you on board to do Phase Two. But no promises at this point.

Consultants: OK. Phase One addresses only the needs of the Sales and Acquisitions area of the General Books Division, right?

Ted: Right.

Consultants: More specifically, we're to analyze the processes and data required to maintain client, item, and source files and to process purchase orders and sales, whether direct -- that is, from your book inventory -- or brokered -- that is, where you act as a broker between a client and a source. Is that right?

Ted: Right. But don't define that so narrowly that you don't see the tie-ins to Communications and Accounting. For example, Accounting is involved in processing purchase orders and sales orders. So you should consult with them as well. Your purpose in this phase is not to 'redo' Accounting and Communications. But, since they're the focus of Phase Two, we need to make sure that any system developed for S&A is compatible with their needs as well.

Consultants: Understood. So we're to consider the interfaces to accounting and communications systems, but not actually fully analyze, design, or implement those systems. For example, if it seemed a good idea to create a purchase order data entry screen for S&A, we should consult with Accounting about their requirements for a PO. But we won't actually create the automated system to update the general ledger, etc.

Ted: You've got it.

Consultants: OK. That's pretty clear. I assume that we're also authorized to address the contact, sales, and other reporting needs of Sales and Acquisitions and to create whatever databases, data entry screens, reports, etc. that are needed for these functions.

Ted: Yes. Basically, your charge is to automate their existing system. Whatever reports are currently created should be created by the new system, unless Jonny indicates that they're not needed. Jonny's a smart guy. I trust him to define what's most important for Sales and Acquisitions, so whatever he says goes. Within reason -- or, more specifically, the budget -- of course.

Consultants: Of course. As long as you've mentioned it, let's turn our attention to the development budget for a minute. $100,000, right? That's to include all hardware, software, data conversion, consulting fees, cabling, and redesign of the facility, if necessary, ...

Ted: Everything, including the cost of training the Sales and Acquisitions staff, writing documentation, etc. You'll also have to identify someone to maintain their systems during operation. Let me see the estimated cost of a part-time systems -slash- network -slash- whatever person in your cost/benefit analysis.

Consultants: Sure thing, Mr. MacCormick.

Ted: By the way, I hope that you noticed the clause in the request about a bonus if the system is completed under budget and/or under schedule?

Consultants: We certainly did. It's our goal to cover development and have something left over for first year production costs.

Ted: Good. Do that and Phase Two is yours. The thing is, General Books has always had a negative cash flow. We fund them out of the profits of other divisions. My real goal is Phase Three, and I'd like to get there as soon as possible. Although I encourage you to come in under budget, I don't want you to skimp. Give them a solid, up-to-date system that will last at least 5 years. I don't want to repeat this effort any time in this century if I can avoid it.

Consultants: Will do. Are there any other constraints on development, other than schedule and budget?

Ted: I'd like to think that we can leverage current investments. The existing PC's aren't so out of date that we can afford to scrap them. I assume that they can be upgraded.

Consultants: Pretty easily. But given that the division only has 4 PC's, incorporating them into the new system won't save a lot of dollars. Nonetheless, a penny saved is a penny earned.

Ted: Ah, that's an adage that goes straight to my Scottish heart.

Consultants: Any other restrictions?

Ted: Keep production costs down. Ideally, one of the existing GB employees would be trained to take over system management. It seems to me that guy in accounting, what's his name?

Consultants: Greg Loi?

Ted: Yeah, that's the one. Isn't he finishing his degree at SUNY? Seems to me he was studying computers. Maybe he'd be able to take responsibility for the system. You know, half-time helping Christina, half-time managing the system.

Consultants: Sounds like a good idea. I'll check into it. One last question, Mr. MacCormick, then we'll get out and let you get to work. In the cost/benefit analysis, ...

Ted: [laughs] Let me guess. You're wondering what the benefits are, given that I've already said that I don't expect to make much money on this. To make things a bit easier for you, consider this. Overtime for Sales and Acquisitions clerks cost me an extra $20,000 last year. I assume that no overtime will be required under the new system. In addition, I'm expecting that, with the new system, General Books should be able to make a profit in the neighborhood of $35,000 to $40,000 by the end of the first full year of production; I'll expect that to increase quite dramatically in the second and third years -- say 25 - 40 percent -- than stabilize at about a 10 percent increase per year. Realistically, we should also be able to cut at least one person from the payroll, but I'm afraid that Ambrose and Jonny would raise a stink. They think they're one big happy family. Problem is, 'Daddy' has to foot the bill. Let me know the feasibility of reducing personnel, anyway. Not that we'll necessarily do it, but, if someone quits, I'd like justification for not hiring anyone to take his or her place.

Consultants: OK. Finally, let me make sure that we're clear on your objectives for the system. Your long-term goal is to increase BookPlace prestige and to use the electronic book trading system as a marketing device. But what are your objectives for the GB internal system?

Ted: Reduce costs, improve employee morale -- Ambrose actually smiled at me in the elevator last week! Exploit Jonny. By that I mean that Jonny is potentially quite a resource for BookPlace. I want him to spend less time pushing paper and more time making sales, promoting BookPlace among the 'literaria,' just generally being an ambassador for BookPlace. So whatever makes Jonny more efficient and effective makes me a happy man. More specifically, I think we should be able to double our transaction volume with the internal system. I've recently learned that the rare book trade can be quite lucrative. Got a look at Swain Galleries' P/L statement last week. They operate at a nice profit. If we can increase our volume, I think we can do the same.

Consultants: So your main objective is to increase productivity, thus generating more sales and increasing revenue.

Ted: Not just revenue. PROFIT! The new system should improve GB's efficiency so that they can finally generate a positive cashflow.

Consultants: Okay. Thanks for your time this morning. We'll keep you informed of our progress. Oh, by the way. We're planning a joint session -- joint application development -- with users, customers, sources, and us consultants. Would you be willing to take a few minutes to kick off the opening session?

Ted: Sure. Just check with my assistant about my schedule. Now, get out of here so I can get some work done.

Interview with Ambrose Yurkel

Ambrose: Good morning. Welcome to our humble abode. I imagine that my office is quite a letdown after spending time in the palatial penthouse suite of our executive leadership.

Consultants: [looks around] Hey, your office has a comfy feel. Not so hard on the eyes as all that glass and polished chrome.

Ambrose: Yes, well, we make do. How can I help you this morning?

Consultants: Just a few questions, Mr. Yurkel. First of all, we need to quantify the expected benefits of the system. Mr. MacCormick noted that one benefit should be a $20,000 savings in overtime charged by Sales and Acquisitions clerks last year.

Ambrose: You must understand that Mr. MacCormick is prone to hyperbole. Our records show that overtime for the whole division was $17,256.81 last year. And much of that was caused by a leaky ceiling that required the clerks to recreate a number of water-damaged files. Typically, overtime runs no more than $14,000 and wouldn't be necessary at all if those poor clerks didn't have a mountain of paper to manage.

Consultants: No one's trying to place blame, Mr. Yurkel. We're just trying to quantify likely savings from implementing the system.

Ambrose: Certainly. Please continue.

Consultants: OK. Let's see. Can you think of any other benefits to be derived from automating Sales and Acquisitions?

Ambrose: How can that man talk benefits? Is it so much to expect that we be brought into the twentieth century? I suppose that our clerks should sit at high desks and maintain their ledgers with quill pens, working by the light of one candle stub --

Consultants: Yes, well .... Mr. MacCormick has also indicated that he expects GB to double its transaction volume and to generate a positive cashflow.

Ambrose: What? Of course .... I knew that there had to be a more direct profit motive behind his ploy. He's always viewed us as the pig's ear. Now he expects us to become a silk purse, a cash cow -- if I may mix my metaphors.

Consultants: Perhaps you can describe your vision of the system. How do you expect it to help General Books achieve its goals?

Ambrose: I expect little and yet am so often disappointed. By the way, I hope that the cost of some new furnishing was included in the budget. We could all use some new office furniture. Each computer should be situated on one of those nice L-shaped tables. You know the kind I mean.

Consultants: Certainly, Mr. Yurkel. Can you describe the specific aspects of Sales and Acquisitions functions that you feel are most critical to your division's profitability?

Ambrose: As Mr. MacCormick has no doubt informed you, General Books and profitability have an avoidance-avoidance relationship. We haven't been profitable since Mr. MacCormick reassigned inspirational and self-help books to the Popular Books division. He takes away our good seed and yet expects us to raise a productive crop.

Consultants: Yes, it's no doubt a difficult situation. Thank you for your time this morning, Mr. Yurkel. I expect that Mr. Wassenaar can fill us in on the details about Sales and Acquisitions. May we speak with you again as the need arises?

Ambrose: Certainly. I'd be pleased to help in any way I can. [rises and shakes hands all around]

Interview with Jonathan Wassenaar

Jonathan: Come in, come in. I hope that you'll be comfortable on these chairs. They're quite a mish-mash of styles, but I've always found that a wooden chair equipped with a nice seat cushion is the perfect work chair. Don't you agree?

Consultants: These are lovely cushions. I'd bet they were hand-embroidered.

Jonathan: And you'd be right! My wife is a demon with a needle and thread. Over the years, she's made cushions for all our chairs. Adds a nice touch, I think. I like an office to look as much like a home as possible. Especially since I so often spend more time here than I do at home! [laughs]

Consultants: You seem in a chipper mood this afternoon, Mr. Wassenaar.

Jonathan: Call me Jonathan. Yes, I'm very pleased -- excited, even -- that we're about to embark on this project. I don't know much about computers, but I'd like to learn. Though you must keep me in line. I don't want to waste your valuable time educating me when you could be doing less arduous work! [laughs]

Consultants: No problem. We're always pleased to encounter users with your level of enthusiasm. You should feel free to ask questions whenever the urge strikes.

We're here to help you become more efficient and effective. A little knowledge about information systems will make you a more informed, productive user.

Jonathan: Good. That's settled, then. How can I help you this afternoon?

Consultants: Well, this is a little touchy, but I feel that it's an issue that needs to be addressed. We've noticed a certain degree of resentment and ill-will concerning the project and want your take on the situation.

Jonathan: Would I be correct in assuming that you've recently talked with Ambrose?

Consultants: Well ....

Jonathan: No need to answer. You must understand that Ambrose has reasons for the way he feels. However, I don't see that his attitude will jeopardize this project. He knows that this is a good thing for General Books, and, although he may not contribute much, he will not do anything to impede your efforts.

Consultants: Then you feel that most GB employees support the project? You support it yourself?

Jonathan: Certainly. Let me draw an analogy. We bibliophiles and the books we strive to preserve are an endangered species. Just like the elephants, cheetah, wildebeest, and other wonderful creatures on the plains of Africa. If the only way to preserve these creatures is to make them too valuable in terms of tourism dollars, then I say "Bravo! Bring on the tourists." When the tourism dollars come rolling in, the rhino will be valued for its whole being, not just its horn. No one will pay to see the rhino horn. Ah, but to see the whole rhino on its home turf? Now that's a sight worth paying for. It's the same with us bibliophiles. If Mr. MacCormick wants to exploit us in his marketing scheme and to generate prestige for BookPlace, then I say "Do it!" I'm not one to bite the hand that feeds me.

Source and Book Selection Process

Consultants: Glad to hear it. We were feeling a little deflated, but your enthusiasm has done a lot to 'inflate' us again. Now, to work! The general overview of Sales and Acquisitions processes that BookPlace provided in the request for proposal provided little detail on how you actually select sources and books. Yet it seems that you must have a selection process since there are many hundreds of sources, of which you use a limited few, and many thousands of books, of which you buy a very few and list in your newsletter as 'best buys' just a few dozen each month. Could you describe how you select your sources and books?

Jonathan: Now that's a difficult question. I've been doing this so long that I have trouble expressing why I do what I do. Let me see. First of all, sources. You're right in saying that there are literally hundreds of rare book dealers. But you must remember that many of these are 'small potatoes' so to speak. Given the limited amount of time that I can devote to studying rare book catalogs and newsletters, I have to focus on the best dealers -- the ones with the best, widest reputation who are most likely to come into possession of really fine works. That's not to say that the small dealers don't ever come up with some really desirable finds. But, until the electronic book trading system is in place, I simply can't study all the sources. So, for now, I've selected the biggest, most reputable dealers.

Consultants: Does your source list change from time to time? In other words, do you add or delete sources?

Jonathan: Certainly. Management changes, and suddenly a reputable dealer is less reputable, if you know what I mean. Recently, one of my best sources was bought out by a syndicate. Well, I can tell you that the new management was more interested in making money than preserving books. They purchased a stolen work, resold it at an exorbitant profit to an innocent client, and caused no end of trouble for everyone involved. I dropped that source from my list immediately. On the other hand, occasionally a small dealer earns an outstanding reputation for honest dealing and an eye for quality. That's the kind of dealer who gets added to the list.

Consultants: Our information shows that only name, address, phone number, contact, and similar data is maintained about each source. Yet it seems that you know a lot more about your sources than your data shows. How is this other information maintained?

Jonathan: [points to his forehead] Right here, I'm afraid. I've wanted to get this information down on paper for some time. I'm grooming one of our clerks, Tran Pham -- you may have met her, to take my place someday. So I try, when time allows, to convey this information to her. But it would be a good idea to have some kind of permanent record.

Consultants: And it's information that could prove valuable in the electronic book trading system. Buyers would be especially interested in the trustworthiness of dealers. But that's work for another day, I guess. We need to focus on Phase One. But please be sure to think of it for Phase Three.

Jonathan: I'm not so sure that it's inappropriate to discuss it now. Would it be feasible to include comments about each source in the source file?

Consultants: Certainly. It just means adding additional fields to the records. Is that something that you'd like us to do?

Jonathan: Well, yes, if it's not too expensive and if Mr. MacCormick approves.

Consultants: Mr. MacCormick has indicated that we're to -- let me check my notes -- 'whatever makes Jonny more efficient and effective makes me a happy man.' So, certainly, we can expand the source data to include comments on trustworthiness, etc.

Jonathan: Good. Then if I'm hit by one of those erratic garment trucks, all will not be lost. Though, caution. This data will need to be updated regularly. And I don't know about the legal issues in disseminating it over the book trading system. You'll have to consult with BookPlace attorneys when the time comes.

Consultants: We'll make a note of that. But for internal use there should be no problem. What about your book selection process?

Jonathan: To some degree, I select books that are classics, guaranteed to appreciate in value. These are the ones that I buy for our inventory. By the way, Mr. MacCormick may find that GB is not so very dollar poor if he views our inventory as an investment. One of our rarest items has appreciated 700 percent in the last 7 years.

On the other hand, I also select books on the basis of what I know our clients want. For example, one of our very wealthy clients collects children's books. She demands high quality and is willing to pay for it. And, what's important to me, she also takes exquisite care of her treasures: stores them in the right environmental conditions, yet also takes them out now and then to feast her eyes on them. A real bibliophile. So I'm always scouting the book lists for rare finds that will get her interest.

Most of my selections are chosen to fill client requests. A client calls and asks me to locate a first edition Shakespeare and I set to work. Another client calls and asks me to locate a Gutenberg bible and I say, "Good luck! No one's selling!" But wouldn't I love to make that sale!

Consultants: Wouldn't we all! Do you maintain a file of client requests, things that you'll need to search for?

Search and Brokered Sales Processes

Jonathan: Yes. We have a tray labeled Search Required. Whenever we receive a client inquiry or order that requires a search, we put it in that tray. Then Tran or I conduct the search, reviewing our book card files and catalogs and contacting sources.

Consultants: Then those items will involve a brokered sale. How do you process a brokered sale? I mean, when you act as an intermediary for a client and obtain a book from another source. How do you charge the client for your services?

Jonathan: In general, we charge a 15 percent finder's fee -- a commission, if you will -- on each brokered item. So, for example, Source A lists Book B for $5,000. When we sell it to a client, we charge $5,750 to cover our 15 percent finder's fee. Now, you must understand that these are items that clients have specifically requested us to locate. Thus, we are performing a personal service for a specific client.

Consultants: Good. That clears up a major source of confusion. OK. Let's move on to discuss how you process sales and acquisitions.

Sales and Acquisitions Processes

Jonathan: Ah, now that's pretty straightforward, I'm sure you'll be glad to hear. First, sales. Because we deal in items that are one of a kind, our clients want to be assured that an item is available and that we agree to sell it to them before they formally submit a purchase order. This is usually arranged through a phone call from long-standing clients or through a telegram or fax from less established clients. We call such a message an Intent to Purchase.

Next, one of two things usually occurs. First, clients that are in the business themselves -- that is, other book dealers, libraries, museums -- submit a purchase order stating their name, address, and so on, and describing the items they want to purchase. These clients -- if they've done business with us before -- have a client account and are billed for the items at the time the order is shipped. As you might expect, non-profit organizations, such as public libraries and public museums, are tax exempt; so they also provide their tax exemption number.

Second, clients who are private collectors usually just call or fax an Intent to Purchase to indicate the items they'd like to purchase. Again, we will agree to process a sale for clients who have an account just on the basis of a phone call. For non-account clients, we require a credit card number or check to process the sale. Whatever the case, private collectors must pay in full -- either on account or by credit card -- before we ship the items to them.

Consultants: OK, that's pretty clear. How do the clients indicate the items they want to purchase. I mean, for example, your recent list includes two *Absalom, Absalom!*

items, two *A Farewell to Arms*.... Do the clients have to repeat all the information about an item to identify it uniquely? You know, binding, condition, etc.?

Jonathan: If the items of interest are from a list in our newsletter, a client describes the book by its book ID and repeats the title, source, and current price as verification. [see Figure BP-7 for sample book IDs].

Consultants: Who assigns these unique IDs to the books?

Jonathan: That's done by a central clearinghouse, if you will. The International Federation of Rare Book Dealers provides several services to its members, including assigning a unique ID as each book first hits the market and authenticating books upon member request.

Consultants: Authenticating books?

Jonathan: Yes, a process similar, I suppose, to authenticating works of art. When we sell a very rare, very expensive item, the client often requests a certificate of authenticity. As a member of the International Federation, we can use their authentication service, paying a fee that typically runs between one and three percent of the book's value. If we are brokering a sale, the source pays the fee and provides the certificate; if we are selling an item from our inventory, we pay the fee and provide the certificate. In both cases, we usually add the authentication fee to the selling price of the book.

Consultants: OK, we've noted that. Now your acquisition process. I imagine that it is probably just the reverse of how you process sales to other book dealers.

Jonathan: Pretty much. To acquire an item, I call the source, we agree on the price and terms, and then I fax an Intent to Purchase to the source. Next, I complete a purchase order form in triplicate. If the item's price is greater than $2,000, I forward the purchase order to Ambrose for approval, though, as a rule, Ambrose is just a 'rubber stamp.' He has never not approved one of my purchases. Once approved, the purchase order goes to Accounting, where, as I understand it, they cut a check and send the check and a copy of the purchase order to the source. There may be more to it than that, but I don't really know the details.

Consultants: We'll check with accounting to get the full story. For now, let's move on to the personalized letters you send to selected clients.

Customized Client Letters, Newsletter Lists, and Item Tracking

Jonathan: Another labor-intensive task. I'll give you my perspective on the process. But again, you'll need to speak with my clerks and Communications to get the full story. Basically, I ask one of the clerks to pull the files of all clients who have made a purchase in the previous quarter. Assuming our client records are up-to-date, the items sold to each client should be listed on his or her client record [see Figure BP-2]. My clerks review the client records and prepare notecards indicating the client's address and recent purchases. Then I hand-write 3 or 4 different form letters, sometimes more. If time allows, I like to compose a form letter for each of our subject divisions: Child lit, science fiction, etc. Then the appropriate letter is sent to the client. That's it. I can't tell you anything specific about the 'behind-the-scenes' processes that occur.

Consultants: That's fine. We'll get that information from the clerks. One last area to cover: the newsletter. I assume that you hand-write the text of the letter. But how do you compile the list of books for your newsletter?

Jonathan: We maintain a 3 x 5 card file of non-inventory items of interest. Usually, I just highlight listings in the catalogs, and one of my clerks prepares a card for each of the highlighted entries. As I compose each newsletter, I select items from the non-inventory book card file to include in the list. The problem is that by the time I'm ready to include an item in our newsletter, it may already be sold. It's very difficult to keep our files up to date. That's why I'm so eager about this book-trading system that Ted has envisioned.

Consultants: I can believe it after seeing your book card files. Reminded me of the card catalogue at a public library. I opened a few drawers and scanned a few cards. One of them had been created June 14, 1983! That's over ten years ago! Your book records -- though neatly alphabetized -- must be a total mess!

Jonathan: A mess, but not a *total* mess. In some businesses, a record that had not been updated in ten years would be useless. In our business, it isn't. For example, by contacting the source listed on that old card, I could begin tracing the current whereabouts of the item. Now, I wouldn't bother to do so if it was an inexpensive book, but, if it was very rare and promised a large commission if located, I'd trace it through as many dealers as necessary.

Consultants: Do you mean to say that you can call the source -- assuming it's still in business -- and the source will tell you who purchased the item? That's hard to believe....

Jonathan: We in the trade are a close-knit community. As a service to each other, we all maintain records -- seemingly forever -- about all the items we sell from our inventory and freely share that information within the community. Of course, there are occasions when the buyer requests anonymity, but, for the most part, we are free to inform reputable dealers about the client who purchased an item through us.

Consultants: But that implies that the system will need to maintain records on book inventory items even after you've sold them. Normally, when an item is removed from inventory, its record is deleted. But, in your case, the record will need to be maintained.

Jonathan: That's right. You may have noticed as you scanned the book files that we maintain separate drawers for books that are in our inventory, books that were sold through us, and books available through other dealers. If you scanned those Books Sold drawers, you'd find many battered cards written by Irvin Yurkel and his son Samuel.

Consultants: Whew.... This changes our estimate of how much secondary storage you're going to need. We had envisioned needing to maintain items currently in inventory and on your best buys list; we didn't anticipate having to convert *every* card in your book drawers to disk! There must have been 200,000 cards going back about 50 years in those drawers .... And we've got to design the system to maintain historical data on all those books ....

Jonathan: Don't look so flabbergasted. It's not quite as bad as it seems. Before we begin the data conversion process, I and Tran will examine the book cards and purge those that are no longer of any interest. That should eliminate about 25 percent of the cards.

Consultants: That's still a lot of historical data. I wonder if it wouldn't be best to maintain that data on CD-ROM .... I'm beginning to think that Mr. MacCormick didn't miss the mark by much when he budgeted $100,000 and six months to this phase of the project. It's going to cost a small mint and take a lot of hours just to convert all that data from cards to magnetic storage. I'm beginning to doubt that we'll be able to complete this project ahead of schedule ....

<u>Interview with Tran Pham, Lisa Edmonds, and Jaime Guzman</u>

Consultants: Good morning. Thanks for meeting with us today. We'll try not to take up too much of your time. We just need to fill in some of the blanks in the information that Jonathan has given us.

Tran: Those 'blanks' may be more involved than you bargained for. Jonathan is a 'big picture' guy. Our job is to handle the thousand detailed pieces required to form that big picture.

Consultants: I can believe it. We almost missed a very important detail when we discussed book records with Jonathan. To be honest, we're somewhat afraid of discovering another sink hole as we learn the details from you.

Tran: Sink hole? What do you mean, 'sink hole'?

Consultants: A sink hole is a soft area of ground that, on the surface, looks fine. But all of a sudden, it gives way, taking cars and houses with it. Our hopes of completing this project on time and within budget were swallowed up by a sink hole called your book files.

Tran: It's not as bad as it seems. Back in the 60's when Jonathan came to work here, he started using color-coded cards to categorize each book as Child Lit, Science Fiction, etc. I've worked here for eight years. About five years ago, when Jonathan started training me to do searches, I decided that searching through all the catalogs to identify books in a particular category was too time consuming. So I started creating a category index, kind of the equivalent of a card catalog, where every book is cross-indexed by author, title, and category. So there are a lot of duplicate cards in the files.

Consultants: I'm glad to hear it! Maybe there's hope after all.

Lisa: [Laughs] Maybe, maybe not. You probably haven't seen the book drawers in the basement.

Consultants: [Groans] Don't tell me ....

Tran: Ignore her. She's giving you a bad time. Yes, there are additional book records in the basement, but very few of those items are of interest any more. Jonathan and I sorted through the book cards a long time ago, pulling items of marginal interest and moving them to the basement. I assume that you won't need to include those records in the new system. I mean, I've had occasion to retrieve a card from those files only twice in five years. So they're of little value. Also, once we run through the records again, we can probably eliminate a lot of the cards in the files up here.

Consultants: I don't know how you keep your sanity trying to maintain all this data manually. I mean, the idea that you were creating duplicate cards to be able to track books by author, title, or category .... With a database, you can maintain one record and retrieve the books by searching on *any* field.

Lisa: I know, believe me, I know. The worst search I ever had to do involved locating books -- any book -- signed by Theodore Roosevelt.

Book Data Definition and Search Process

Consultants: Assuming we set up the database correctly, you should be able to search the files in a matter of minutes and receive a list of items fitting your search criteria. But, to do that, we need to know what those search criteria are. What kinds of searches do you commonly perform?

Tran: Well, as I mentioned, we have to search by category. Some clients buy books of a certain genre; they're less concerned about the author or title of a book, but very concerned that it be a rare item in that genre.

Consultants: So you need for each book to be assigned to a genre category. What categories do you use?

Lisa: That task's been assigned to me. I brought an excerpt of a report [see Figure BP-11] I used to generate for Jonathan.

Author	Title	Category
Anderson	Winesburg, Ohio	3
Faulkner	Absalom, Absalom!	3
Faulkner	Light in August	3
Hawthorne	The Scarlet Letter	1
Heinlein	The Green Hills of Earth	6
Hemingway	A Farewell to Arms	3
Hemingway	The Old Man and the Sea	3
Jackson	Haunting of Hill House	4
Kerouac	On the Road	4
Melville	Moby Dick	1
K.A. Porter	Pale Horse, Pale Rider	3
B. Potter	Jemima Puddle-Duck	5
Whitman	Leaves of Grass	2

1 = 19th Century American Fiction   2 = 19th Century American Poetry
3 = 20th Century American Fiction   4 = Modern Fiction
5 = Children's Literature   6 = Science Fiction

Figure BP-11: Interest/Genre Categories

Consultants: Are all books you track assigned to a category?

Lisa: Yes, that's part of my job. At first, Jonathan noted the category when he highlighted an item in a book catalog. When I first started doing this, Jonathan wasn't sure I could be trusted. That's why I used to generate this report. But he trusts my judgment now, so assign the category and no long need to generate this report.

Consultants: Thank heaven .... What other kinds of searches do you need to do?

Tran: The usual ones, you know, by author, title, publisher --

Consultants: Publisher? But the publisher isn't included in the data you maintain for each book?

Tran: Only indirectly. A very few publishers are themselves of interest. If that's the case, we note the publisher in the edition or condition field. The same is true when a book was owned by a famous collector. For example, you may have noticed 'Epstein copy' in the condition field of some books.

Consultants: Should publisher and collector be fields tracked for each book?

Tran: Not really. Only a very few books are of value because of the publisher. From what I understand about databases, you wouldn't want to include a field if 95 percent of the time it was empty. That's true in this case. At least 95 percent of the time, the publisher is not important in and of itself.

Consultants: Are there other miscellaneous attributes that may be of interest?

Tran: Well, as I said, the inscription -- if there is one -- can be of interest. Especially if the book was signed by the author or inscribed to a famous person.

Consultants: From the sounds of it, it may be worthwhile to include an attribute to track this kind of information. For example, we could create a field called MiscData and use it only to indicate the publisher, collector, or inscriber. If the field contained the name of the publisher, its value might be "P-Scribners"; if it contained the name of the collector, it might be "C-Epstein." That way, you could do searches on publisher, collector, or inscriber without having to maintain separate fields for them.

Tran: OK, but what if a particular item was of interest because of all three attributes? Let's say, because it was published by Scribner's, collected by Epstein, and signed by Hemingway?

Consultants: I see what you mean. We'll figure out some way to deal with that situation. I assume it's fairly rare.

Tran: Yes, rare, but very valuable. That book would be worth at least $50,000. So it's data we'd need.

Consultants: OK. Anything else about searches? No? Then let's verify our understanding of the item tracking process. Here's the sample book data we were given [see Figure BP-7]. As you've indicated, we'll need to add a miscellaneous field to this data. Is there any other data you need about each book? Would it make sense to treat books that you actually own separately from books that are available on the market? What I mean is that it seems that you have to maintain more data for a longer period of time on books that are or were in your inventory. Maybe it makes sense to create separate files: one for interesting items available from other sources, one for items in your current inventory, and one for items sold from your inventory.

Jaime: I absolutely agree. As you know, we have to maintain data about books we sell from inventory forever; so it makes sense to treat them separately.

Consultants: OK. Anything else about book tracking? OK, let's talk briefly about how you create the customized client letters.

Customized Client Letters

Consultants: Lisa, I understand that you're primarily responsible for that task. How do you identify clients to receive customized letters?

Lisa: Well, first I have to pull the sales orders (invoices) for the past quarter. These list who bought what item when. I create a 5 x 7 notecard for each client, listing the client's name, address, and items and date purchased. I give these to Jonathan, who notes on the card any special, personalized sentences he wants included in the form letter. He puts each form letter, along with the notecards related to that letter, in a separate folder and gives the folders to the Communications staff. For example, he may write five form letters and personalize one of the five for each client.

Jaime: Recently, Jonathan has asked me to write some of the letters for him. Specifically, to new clients who made sizable first-time purchases, you know, above $1,000. I write a letter for each client, mentioning similar items at similar prices that we currently have available. The idea is to build a relationship with these clients and to encourage repeat business.

Consultants: That suggests that you'll want to be able to sort client orders on the basis of the dollar amount of the order. Is there any need to accumulate sales for each client?

Jaime: Currently, that's done on a short-term basis, whatever period the sales report covers. But it might be useful to be able to accumulate sales by client. Tran, what do you think?

Tran: Hmmm. Well, I know that Jonathan gives special service to long-standing clients or those who purchase expensive items. This is somewhat off the subject of letters, but it may be useful to add a field -- maybe three fields -- to each client record. One to indicate when a client started doing business with us, one for date of last purchase, and one for accumulated sales. Right now, we know who our best customers are because Jonathan and I deal with them regularly. But eventually Jonathan will retire, I may suffer a head injury, whatever, and someone else will need to be able to identify our best clients.

Sales Order Processing

Consultants: Let's talk about sales order processing for a while.

Jaime: That's my primary responsibility. Whenever we receive an Intent to Purchase from a customer -- either via a phone call, letter, or fax, that information is passed on to me. I complete an invoice form in triplicate and forward it to Accounting. I keep a photocopy of the invoice and place it in the Sales Pending tray. When the items are shipped, I pull the source and client folders associated with each invoice, annotate the appropriate records, and refile the folders. I also file the invoice photocopy chronologically in our Sales Orders file.

Consultants: Assuming the purchased items are from your inventory, how do you process shipment?

Jaime: We have a general policy of not shipping the items until payment has been made. If the client has included a check or credit card number to pay for the item with his Intent to Purchase, I forward it along with the Intent to Purchase and invoice form to Accounting. They verify the check or credit card number, process the invoice, and then return the pink copy of the invoice to me. I include it in the package when the items are shipped. Most of our shipments are handled by certified mail, unless the items are particularly valuable, in which case we use couriers such as Federal Express or UPS to hand-deliver the shipments.

Consultants: What if the order has not been prepaid?

Jaime: Some clients have an account with us that allows them to submit a purchase order and to receive the shipment even though payment has not been made. For example, most of the book dealers, museums, and libraries can be invoiced for a purchase; our terms are full payment within thirty days of the invoice date. Other clients who don't have accounts and haven't prepaid the order are invoiced for the order. Accounting sends the client the white and pink copies of the invoice; the client returns payment with the pink copy. When payment is received, we ship the items with the pink invoice indicating that payment has been made. That copy serves as their receipt.

Consultants: Let's see if we can summarize the sales order processing activities. A client contacts you -- that 'client' can be a private collector, another book dealer, or a museum or library -- indicating interest in a particular item. If you have the item in your inventory and if the client has enclosed payment, you complete the invoice form and forward it and the payment check or card number to Accounting. Accounting returns the pink copy to you marked "Paid"; you package the item for shipment, enclosing the invoice copy as a receipt.

If the item is in your inventory and the client is another book dealer or a private collector who has an account with you, Accounting verifies the client's account and gives you the okay to ship the book. Then Accounting sends the pink and white invoice copies to the client to request payment. Whatever happens next is Accounting's responsibility. The client sends payment to Accounting along with the pink copy. Accounting processes the payment. If the client doesn't send payment, Accounting follows up on it. In other words, accounts receivable is Accounting's responsibility.

If the item is in your inventory and the client hasn't prepaid and doesn't have an account with you, you complete the invoice form and forward it to Accounting. They send the white and pink copies to the client. Once the client pays for the order, Accounting forwards you the pink copy marked "Paid", which you enclose when shipping the item to the customer.

OK. Let's examine the case where the item requested is not in your inventory. Again, the process begins by a client indicating interest in an item. If you know where you can obtain the item, you do what?

Tran: Orders that require any kind of search or interaction with other sources are my and Jonathan's responsibility. We search our book files to locate the item; if it's not in our files, we search book catalogs or contact other sources to try to locate the item. Assuming we succeed, we contact the client and inform him of the item's condition and total price, including our broker's fee. If the client agrees to the price, we prepare an invoice, indicating the selling price, which is actually the source's price plus our finder's fee. The rest of the procedure depends on the conditions you stated earlier: prepaid order vs. order on account vs. invoiced order. The only difference is that Accounting sends the source a purchase order and check, and then the source sends us the item by overnight delivery.

Report Generation Process

Consultants: Well, that's pretty clear. We may have to check back with you again later if questions arise. Finally, let's examine the reporting process. Our information shows that you currently generate three reports: Client Contacts, Sales, and Book Inventory.

Tran: That's not entirely accurate. The Sales and Book Inventory reports are prepared by Accounting. They keep all that information in spreadsheets, so it's easier for them to generate reports. We create the Client Contacts report -- or, more specifically, Lisa and Jaime do.

Lisa: The Client Contacts report is fairly easy to create, but very tedious. Every time we have contact with a client -- whether the clients contact us or we contact them -- we note the date, type (phone, letter, newsletter), and a brief comment about the contact on a notecard. These are batched in a tray labeled Client Contacts; as time allows, Jaime and I update the client folder to indicate these contacts. But we retain the notecards because it's easier to create the Client Contacts report from them than to search through all the client folders. To create the report, we sort the cards by client name, grouping contacts with each client and arranging each group in chronological order. Then we type the report from these sorted cards. That's it.

Consultants: Again, if you maintain contacts as part of your database, you'll be able to automatically generate that report. I have the feeling that you're going to love the new system! So much less paper pushing to occupy your time ....

Jaime: You better believe it! I'm thinking of feigning illness for about six months, just because knowing that all this tedium will disappear soon makes putting up with it now so much more difficult.

Consultants: I sympathize. Knowing that report generation will be so much easier, can you think of other reports you'd like to have? I'm sure that you've avoided creating some desired reports because of the work involved.

Tran: I'd appreciate having an easy way to summarize the current status of searches. For example, if we could enter a record for each search, giving the client and item involved and the person conducting the search -- Jonathan or I. Then we could update the record as new information is received. A search can span several weeks, even months if the item is very rare. Sometimes I lose track of which dealers I've contacted and end up contacting them again.

Consultants: So you'd like to maintain data about searches in progress. Each record would have some kind of unique identifier relating it to a client and book. The rest of the record would probably be text describing attempts made to find the item. Is that right?

Tran: Pretty much so. What I mean is that when we receive a client request requiring a search, that request should be posted to the search file, including a notation whether Jonathan or I have been assigned to the search. I'd like to be able to call up all the searches to which I've been assigned. I'd also like to be able to update each search as it progresses, noting the date and dealer contacted and outcome. Maybe talking about this is a waste of breath. Will we be doing searches under the new system?

Consultants: We'll check with Mr. MacCormick. I think we have enough information to design the search file and report if he feels that it's relevant under the new system. Any other reports?

Tran: [Looks at Jaime and Lisa; pauses] Let's wait until after you've talked with Mr. MacCormick. Once we have a better sense of what the final system will look like, then we can consider what reports might be of value.

Consultants: I think a light bulb just came on. You're concerned about being dispensable. You're afraid that the new system is going to cost you your jobs. [pauses] I can't say for certain, but I think you're worried without cause -- at least, about losing your jobs. Why would Mr. MacCormick go to the expense of upgrading your systems and automating your current processes if they aren't going to be used once the electronic book trading system is in place?

Jaime: Stranger things have been known to happen ....

Consultants: OK. Meeting with Mr. MacCormick is top priority. We'll get back to you as soon as possible. And, believe me, I think you'll find that the gloom and doom future you envision is not going to come to pass.

Tran: Let's hope not ....

<u>Confirmation Interview with Ted MacCormick and Jonathan Wassenaar</u>

Consultants: Thanks for meeting with us today. I know you're both very busy, but we feel that the Phase One project may be endangered unless we settle a few issues that have arisen as we interviewed some of the clerks. The concern is --

Ted: OK, OK, cut to the chase! No need to summarize. You made clear what the issues are in your phone call requesting this meeting. Let me go straight to the point. Phase Three will NOT put any of the clerks out of work. The idea is that processing costs for inexpensive books are too high. The network is designed to reduce transaction costs -- and increase the volume of transactions -- for low-cost books. Right, Jonny?

Jonathan: That's my understanding now, yes. But I totally understand the clerks' concern. And I agree with the consulting team that this issue needs to be resolved before we proceed any further in Phase One.

Ted: So -- it's resolved. Right?

Jonathan: Yes and no. We know how things stand, but I think it's important that we issue a memo to the clerks in GB describing the general purpose of the electronic book marketplace.

Ted: You write it and I'll sign it.

Jonathan: OK --

Ted: Good. Then this meeting is adjourned.

Jonathan: Not so fast, Ted! Excuse my outburst. Let's just take two minutes to answer any questions the consulting team has ....

Ted: Jonny, you drive my Type A personality crazy .... Oh, all right. Fire away. But just two minutes!

Consultants: OK ....

Ted: One minute and 45 seconds ....

Consultants: The Phase Three network will be used exclusively to buy and sell low-cost books. What is a "low-cost" book?

Ted: Generally speaking, any book whose processing costs exceed 10 percent of the book's value. I've thought about and decided that the general policy of the network will be that no books valued at more than $1,000 will be listed.

Consultants: So ... the idea is that most book sellers can't make money selling inexpensive books anyway. So nobody really loses by placing these items on the network. And BookPlace gains in the sense that you provide a service to bibliophiles around the world, thus upping your prestige and name recognition. In addition, you can charge a reasonable fee for that service and make a little profit. Is that right?

Ted: Right. We're still working out the details of the fee schedule. But that's outside your scope, so no matter.

Consultants: One last question. How will listing only inexpensive books address the problems dealers have in locating rare, more expensive items?

Ted: You miss the point. What we've described so far is just what the public sees. Any Joe Blow can log on to the network and search for a low-cost item. But only book dealers can see the full book lists of other dealers. See? We'll charge a fee for book dealers to post their catalogs to the network, instead of printing and mailing them. The printing and postage costs they save should justify our charging them something! Anyway, book dealers will be able to conduct searches of the database to locate items. And, for sure, we'll charge a reasonable transaction fee for that service.

Anyway, the details don't really matter at this point. All that matters is that you understand two things. One, we're not trying to put anybody out of work. Two, we're trying to introduce transaction efficiency into what has been a very inefficient business. A lot of the wealthy collectors will still demand the personal service that we provide. But we do a service by making "bibliophilia" -- or whatever the heck you call book collecting -- an accessible hobby or interest for a wider audience of people. Amen. Now get to work!

<u>Interview with Christina Domingo and Greg Loi</u>

Consultants: We understand that you actually produce two reports for Sales and Acquisitions: the sales report and the book inventory report. Can you tell us a little about that process?

Christina: Sure. It's pretty simple. We maintain spreadsheets recording all sales and acquisitions. Let me bring up a sales spreadsheet as an example (see Figure BP-12). Each sales spreadsheet includes sales for a given time period; this one lists sales during April 1995. When we receive invoices from Sales and Acquisitions, we enter the invoice data into the spreadsheet. To generate the report, we just print a formatted spreadsheet -- formatted in the sense that we group sales by type and by client and include a total for each type and a grand total. That's the extent of it.

Consultants: So each sales period actually requires two spreadsheets: one that lists sales sequentially as they occur and one that is formatted as the sales report.

	A	B	C	D	E	F	G	H	I	J	K
1	Sales for April 1996										
2	File name: sales.xls				Paymt Rec'd						
3	**Name**	**Invoice #**	**Inv. Date**	**Payment**	**Appr. code**	**Book ID**	**Author**	**Title**	**PPrice**	**SPrice**	**Cost of Sales**
4	Baker, Marilyn	1173	4/25/96	BA032	5/10/96	POT12075	B.Potter	The Tale of Peter Rabbit	$0.00	$57,500.00	$50,000.00
5	Philberts, Son & Nease	1181	4/26/96	PO#973-431	5/5/96	JAC33333	Jackson	Haunting of Hill House	$45.00	$100.00	$55.00
6	Baker, Marilyn	1185	4/27/96	BA032	5/10/96	POT00222	B.Potter	Jemima Puddle-Duck	$650.00	$765.00	$650.00
7	King, Kenneth	1190	4/28/96	V41829731446	4432	POR99923	KAPorter	Pale Horse, Pale Rider	$80.00	$110.00	$80.00
8	Charles, Frank	1198	4/28/96	MC4081978534	7109	HEIC1321	Heinlein	Stranger in a Strange Land	$700.00	$805.00	$700.00
9		1198	4/28/96	MC4081978534	7109	HEIC0980	Heinlein	The Green Hills of Earth	$50.00	$120.00	$70.00
10	...										
11											
12	PO = Purchase order# related to invoice							Total Sales:		$101,795.00	
13	MC=Mastercard, V=Visa, AX=American Express							Total Cost of Sales:			$73,448.00
14	C = Check/Money order							Adjusted Revenue:			$28,347.00

Figure BP-13: Example of Sales Spreadsheet
(see sales.xls on project disk)

Greg: That's right. I know what you're driving at. Why two spreadsheets? Well, it really isn't a big deal. I mean, all we have to do is copy the sequential spreadsheet and rearrange some of the rows, adding a few formulas to calculate subtotals and totals.

Consultants: Am I correct in understanding that whichever of you has the time is the one who prepares the report?

Christina: Yes. We both understand what's involved, so either of us can do it.

Consultants: OK. How about the book inventory report?

Greg: I usually handle that. As we process purchase orders, I enter the PO data into a spreadsheet. Also, as we receive invoices for books in our inventory, I annotate the appropriate book record to indicate that it has been sold. The Book Inventory Report you were given [see Figure BP-10] was generated directly from my spreadsheet [see bookinv.xls on project disk].

Consultants: Hmmm .... That raises some questions. First, I understand that the S&A clerks annotate the book card to indicate information about its being sold.

Greg: I know. But that information could be lost -- notecards have a tendency to disappear. Besides, I need to update the book inventory spreadsheet anyway, so adding info about the sale is no big deal.

Consultants: O.K. Here's another question. How do you know the current price of a book in inventory?

Greg: Jonathan updates that information occasionally. When he receives a Book Inventory Report and notices that a current asking price for an item is outdated, he edits the price and returns the report to me. Then I update the spreadsheet.

Consultants: We didn't know that. No one else has mentioned it.

Greg: They probably figured that was our job and that we'd explain it.

Consultants: Then it seems that you actually perform some of the book tracking activities for Sales and Acquisitions. Recording sold book information and tracking prices are integral aspects of the item tracking process that need to be addressed in Phase One. I'm just glad you thought to mention them.

Christina: We've always considered it part of the purchasing process.

Consultants: I guess there's an overlap in responsibility. How about payments? How do your process payments from clients and to sources? For example, the Sales and Acquisitions clerks have informed us that they complete the invoice, but that you verify account numbers and so on.

Christina: That's correct. When we receive the invoice, the client's credit card number has already been entered on the invoice. We contact the appropriate bank to authorize the client's credit card purchase. For account sales from museums, libraries, and other dealers, we receive the purchase order with the invoice forms from S&A. For account sales from most of our private collectors, we consult our client account records and verify that the purchase does not exceed the client's credit limit. Then, we annotate the invoice, adding the PO reference number and date, or writing "Approved," the credit card authorization code, and our initials beside the account number, whichever is appropriate.

Consultants: What kind of information do you maintain about client accounts?

Christina: Name, address, account number, that sort of thing ... plus the client's credit limit and current balance, if any. As you probably already know, all accounts must be paid in full within 30 days of the invoice date. We also maintain records of all transactions on our accounts, as you saw in the sales spreadsheet.

Consultants: OK, that pretty well covers payments from clients. How do you process payments when you purchase books?

Christina: When we receive a BookPlace purchase order -- either directly from Jonathan or approved by Ambrose, we usually cut a check immediately and send the check and purchase order to the dealer. Only for large purchases, those more than $5,000, do we put the order on our account with the dealer -- we have accounts with most of the dealers we commonly interact with. In addition, each year we have a budget for acquiring books -- What's wrong? You seem to have lost interest.

Consultants: Hmmm .... Sorry. It's just that I'm not sure that all this is territory we need to cover in Phase One. It's good to know that we need to maintain BookPlace's account number with the source as an attribute of the source. But I guess that, for the time being, you and Communications are to be treated basically as 'black boxes' that receive inputs from and generate outputs to Sales and Acquisitions. Given that premise, I think we've got all the information we need for Phase One: we understand how you generate the Sales and Book Inventory reports and, in general, how you interface with S&A, processing invoices and purchase orders. If we are retained for Phase Two, we'll investigate your accounting systems fully.

Interview with Farah Faisal and Sam Rao

Consultants: I think we have a pretty clear idea of the client letter and newsletter processes. We have just a few questions about the book inventory spreadsheet you use in generating the newsletters.

Farah: That's Sam's territory.

Sam: It's no big deal. All I do is get a copy of Greg's book inventory spreadsheet and then import it into Word. I have to edit out a few of the columns. You know, purchase date and purchase price, for example. No sense letting our customers know how much we actually paid for a book! That's it.

Consultants: What? No sink holes hiding here? What a relief!

Interview with Bobbi Jo Murphy and Katrina Rhu

Consultants: I understand that you have created a mail-merge file of client names and addresses. Can you tell me a little about this file?

Bobbi Jo: Sure thing. We got sick and tired of typing a zillion mailing labels every time Jonathan wanted to send out a newsletter or customized letters. So we took things in our own hands and used Word to create a labels file. Ain't that right, Katy?

Katrina: Yes. We used Word's database capability to create a file of client addresses that we can reuse each time labels are needed. We also use the mail-merge facility to create Jonathan's customized letters --

Bobbi Jo: Hey, why tell when we can show! Just give me a minute and I'll bring up the file. Whirl...whirl.... These 386's are sooooo slow.... OK, here we are. [see labels.doc on the project disk]

Bobbi Jo: Ick! Looks kinda ugly when I bring it up this way, but it gives you the idea anyway. It's really nothin' special. About the only things to notice are that we included a separate field for title -- Mr., Mrs., Mme, Herr, whatever -- two fields for address, in case we need to indicate a company or suite number or something, and then also a field for country since Jonathan does so much business with foreigners.

Katrina: "Foreigners!" Bobbi Jo, you can be so provincial sometimes.

Bobbi Jo: Well, that's what they are. "Foreign" to the Sovereign State of Texas, United States of America, anyway. I didn't mean any harm, Katy. Let me *emend* my statement -- as Jonathan and Ambrose would say: "Jonathan has such a great *volume* of *correspondence* with a *plethora* of people all over this *populous* planet." Is that better?

Katrina: Much. Thank you.

Consultants: Well, anyway. It seems likely that you have saved not only yourselves but also us developers a lot of time. I was dreading having to enter 2500 names and addresses for the client file. We should be able to import your labels file into the Sales and Acquisitions database. Assuming, of course, that your data is reasonably up-to-date.

Katrina: I can assure you that it is. Bobbi Jo or I update the file whenever Sales and Acquisitions gets a newsletter or letter returned. All correspondence has a note by our return address saying 'Address correction requested.' So we enter corrections directly from the post office address correction label -- or from notes that Sales and Acquisitions passes on to us indicating a change in client address.

Consultants: That would maintain existing client information. But what about new clients? How do you receive their address information?

Bobbi Jo: Through a great leap of intelligent detective work! I mean, really, go figure. If we see a name in one of the Sales and Acquisitions letters that isn't in our labels file, we figure "Gee whiz. Guess this must be a new one...." and add it to our labels file. Aren't we amazingly intelligent and perceptive?

Consultants: I guess. But that raises another question. How do you know when to drop a client from the client list? Don't you ever 'lose' clients?

Katrina: Yes. But mostly because they've moved and haven't contacted us with a change of address and haven't left a forwarding address with the post office. That's especially common with overseas correspondence. Oftentimes, undeliverable correspondence is returned to us with "Return to Sender" or the equivalent stamped on the envelope. If that happens, we remove the client's data from our labels file.

Consultants: Sounds reasonable. Thanks for your help. We look forward to talking with you again, assuming we get the Phase Two project.

Name ______________________________

Exercise 1: Goal Analysis

Analyze how well the current system meets the SQ and OR goals; then, indicate what actions you will take to ensure that, as you develop the new system, all of the system development goals - SQ, PM, and OR - are achieved.

Goal	Case Problem Issue(s)	Action(s) to Meet Goal
SQ- Functionality		
SQ- Maintainability		
SQ- Flexibility		
PM- Timeliness		
PM- Cost		
PM- Client Commitment		
OR- Efficiency		
OR- Effectiveness		
OR- Competitive Advantage		

Name ______________________________

Exercise 2: User Analysis

Analyze the system's users by

1. Classifying each user as an end-user, user-manager, or user-sponsor.
2. Assessing the commitment of each user to the systems development project.
3. Indicating how each user can be involved in the project: e.g., do some users have computer skills, business expertise, or other qualities that you can leverage as you develop the system?

End-users:

User-manager:

User-sponsor:

Name ______________________________

Exercise 3: Function/Component Matrix

Analyze the functions and components of the current system by completing a Function/Component Matrix. Do not make up unspecified parts of the system; just enter "not specified" if no information about a particular function or component is provided in the case.

	INPUT	PROCESSING	OUTPUT	STORAGE	CONTROL
PEOPLE					
PROCEDURES					
DATA					
SOFTWARE					
HARDWARE					

Team Member Names ____

Exercise 4: Workflow Diagrams & Affinity Diagram
(page 1 of 2)

1. Meet with your project team to assign each team member to draw a **user-level workflow diagram** based on an interview with one or more of the users. List each user-level workflow diagram and its creator in the space below:

User-Level WFD Title	Team Member Who Created WFD

2. As a group, compile the user-level diagrams to **create a combined user-level workflow diagram and an organization-level workflow diagram** of the current system.
3. Clearly label your diagrams (providing a title for each, e.g., "User-level Workflow Diagram of Interview with ______________") and attach them to this form.
4. On your combined user-level diagram, indicate the **system boundary**.
5. Based on your organization-level workflow diagram, **identify the major processes of the system**.

Major Processes of ____
(system name)

____

____

____

____

____

____

____

____

Team Member Names ____

Exercise 4: Workflow Diagrams & Affinity Diagram

(page 2 of 2)

6. **Analyze the combined user-level diagram** (a) to identify the internal and external entities of the current system and (b) to make a preliminary list of the system's workflows. **Relate each internal entity to the workflow(s) it creates (C) or uses (U) in the following matrix.**

Internal Entity ⇒ **Workflow ⇓**									

Team Member Names ____

Exercise 5: Enterprise Object Model

1. Examine your workflow diagrams and affinity diagram from Exercise 4 to identify candidate object classes.
2. Then, construct an enterprise object model grouping the object classes into subjects by process. Use the major processes you identified in Exercise 4, question 5 as the subjects of your enterprise object model.

Name ______________________________

Exercise 6: Logical DFD of the Existing System

1. Meet briefly with your team to confirm the major processes of the existing system, using your answer to Exercise 4, question 5 as your starting point.
2. *Working together*, specify the Level 0 DFD of the existing system.
3. Then, assign one of these Level 0 processes to each team member. *Working individually*, create a balanced Level 1 (and lower levels, if necessary) DFD to document the IPOSC functions of your assigned process in greater detail.
4. DFDs can be used to determine the input, processing, output, and storage functions of an information system. Analyze the child DFD(s) you individually created to state the IPOSC requirements of your segment of the system.
5. Attach your Level 1 (and lower) DFDs to this form. Submit all team member exercises together with your team's Level 0 DFD.

Child DFD(s) of Process ______________________________
(Process ID number and process description label)

Input Functions -- The system must accept the following inputs:

Processing Functions -- The system must perform the following processes:

Output Functions -- The system must generate the following outputs:

Storage Functions -- The system must maintain the following data:

Control Functions -- The system must enforce the following controls:

Name ______________________________

Exercise 7: Critical Success Factor Analysis

Consider the five sources of critical success factors -- industry, competitive, environmental, temporal, and managerial -- to state the organizational, functional area, and information system CSFs of the system you are investigating.

Organizational CSFs (i.e., for the organization as a whole)

Business Area CSFs (i.e., for the particular division or business unit being studied)

Process CSFs (i.e., for the particular process being studied)

Name ______________________________

Exercise 8: System Objectives, Project Constraints and Scope

1. System Objectives. State the business goals that the organization hopes to achieve by implementing the new system.

2. Project Constraints. State the constraints on the development project. Briefly discuss the reasonableness of these constraints; explain your strategy for completing the project within the stated constraints.

3. Project Scope. Define the boundaries of the new system. Pay particular attention to stating clearly the workflows, functional areas, or processes encompassed by the proposed system.

Name ______________________________

Exercise 9: Project Risk Evaluation

1. Evaluate the risk of the project. To receive full credit for this exercise, use the Comments section to explain your ratings.

Project Risk Evaluation		
Factors Affecting Project Risk	**Rating**	**Comments**
1. Characteristics of the organization a. Has stable, well defined objectives?		
b. Is guided by an information systems plan?		
c. Proposed system fits plan & addresses organizational objectives?		
2. Characteristics of the info. system a. Model available/ clear requirements?		
b. Automates routine, structured procedures?		
c. Affects only one business area? (no cross-functional or interorganizational links?)		
d. Can be completed in < 1 year?		
e. Uses stable, proven technology?		
3. Characteristics of the developers a. Are experienced in chosen development methodology?		
b. Are skilled at determining functional requirements?		
c. Are familiar with technology and information architecture?		
4. Characteristics of the users a. Have business area experience?		
b. Have development experience?		
c. Are committed to the project?		
Total Points		

2. Briefly describe strategies to minimize the risk in this project.

Name: ______________________________

Exercise 10: Object Relationship Model

Meet briefly with your project team members to assign one of the system's sample source documents or reports to each member. Create an object relationship model of the object classes in your assigned document/report, taking into consideration any additional data requirements identified through analysis of the user interviews. In your model, indicate the identifier of each object class (if any), the class's attributes, and the cardinality of each object relationship.

Object Relationship Model for ______________________________
Document/report

<u>Problems and Questions</u>. In the space below, list any problems or questions you encountered or assumptions you made about business rules as you created your object relationship model.

Name: ______________________________

Exercise 11: Physical Database Design

1. Attach a copy of the ORM you created for Exercise 10. Then define the required tables by creating the relations indicated by your ORM. Underline the primary key of each relation; indicate foreign keys using a squiggle-underline. Use both an underline and a squiggle-underline to indicate a data element that is both a primary key and a foreign key.

2. Should you segment or aggregate any of the object classes or attributes in your relations? If so, redefine the appropriate relations here, and explain your justification for segmenting or aggregating.

3. Use a database package to create the table structures defined in your relations. Then use your DBMS to print a data dictionary for one of your files. (See Appendix B for help performing this task using Microsoft Access.) Attach your printout to this form.

Team Member Names ______________________________

Exercise 12: System Behavior Design--Physical DFDs

(page 1 of 2)

1. As a team, create a Level 0 physical DFD of the new system. The Level 0 logical DFD you created in Exercise 6 is a good starting point for this task.
2. Assign each member to explode one or more of the new Level 0 processes in a Level 1 (and lower, if necessary) physical DFD.

Level 1 Physical DFD	Team Member Who Created

3. As a team, partition your DFDs into programs, number each partition (e.g., Partition 1, Partition 2, ...), and decide how each program should be implemented (batch or real-time processing). In the space below, justify your choice for each partition.

Partition Number	Batch or Real-Time?	Justification
1		
2		
3		
4		

Team Member Names ______________________________

Exercise 12: System Behavior Design--Physical DFDs
(page 2 of 2)

Partition Number	Batch or Real-Time?	Justification
5		
6		
7		
8		
9		
10		
11		
12		

4. **Submitting Your Work.** Each individual should compile his/her child DFDs and staple them together as a unit. Collect all the individual units and paper clip to this form along with the team's Level 0 physical DFD.

Name ______________________________

Exercise 13: System Behavior Design--Program Structure Charts

1. Create an overview program structure chart indicating the major programs in your system.
2. Assign each member to transform his or her Level 1 (and lower) DFD into a program structure chart; in other words, decompose one of the high-level modules in your overview program structure chart.
3. Working individually, in the space below, write structured text process descriptions for two modules in your individual program structure chart.
4. **Submitting Your Work.** Collect each individual's exercise and paper clip them together along with the team's overview program structure chart.

Name ________________________________

Exercise 14: System Behavior Design--Use Cases

1. As a team, identify the use cases of the new system. Assign one or more use cases to each member.
2. Working individually, write your assigned use case(s) in the space below.
3. Underline use-case phrases that require system behaviors.
4. Define methods to perform these behaviors and assign each to the appropriate object class (i.e., create an object relationship model segment that also specifies object methods).
5. Attach your object relationship model segment to this form.
6. In the space provided below, write structured text process descriptions for two of your object methods.

Team Member Names ____

Exercise 15: Dialog Flow Diagram

1. As a team, examine your physical DFDs, use cases, and/or program structure charts to identify the screens your system will require.
2. Then, on a separate sheet of paper, create a dialog flow diagram indicating each screen and the flow of control between screens. Attach your dialog flow diagram to this form.
3. In the table below, assign a number and name to each screen and briefly explain its purpose.

Screen Number & Name	Purpose

Team Member Names ____

Exercise 16: Interface Design

1. Use whatever tool you have available (e.g., Microsoft Access's Reports or Forms feature; a word processing, graphics, or spreadsheet program; plain old pencil and paper!) to design one of each of the following user interfaces: a printed report, a data entry screen, and a menu screen. Attach your screen design(s) to this form.

2. Evaluate your interfaces using the criteria of good user interface design.

   Printed Report:

   Data Entry Screen:

   Menu Screen:

Name ________________________________

Exercise 17: System Flowchart

1. Meet briefly with your team to identify the major subsystems of your proposed system and to assign one of these subsystems to each team member.
2. Then, working individually, create a job-level system flowchart to document the hardware, programs, and data files for your assigned subsystem.

Job-Level System Flowchart for ________________________________
(subsystem documented)

Team Members Names: ____

Exercise 18: Hardware & Software Specifications

(page 1 of 2)

1. **Hardware Specifications.** Identify the key hardware devices that will be required. For each device, provide a general indication of the number of units required, the performance levels expected (e.g., processor type, MHz, RAM, disk capacity, print speed and quality), and, where appropriate, any unique specifications that may be needed.

Hardware Devices	**# Units**	**Performance Level**	**Unique Specifications**
Processor/ Computer			
Secondary Storage			
Peripherals			
Telecommuni-cations			

Team Members Names: ____

Exercise 18: Hardware & Software Specifications

(page 2 of 2)

II. Software Specifications. Assume that the organization has decided to purchase an application package as the software component of the new system. Review your IPOSC requirements and other exercises or documentation to identify criteria to use in your evaluation of alternative software packages. Then assign a weight (from 1.0=least important to 5.0= most important) to each criterion. In the space provided, briefly justify your weight for each criterion.

Criterion	Weight	Justification
1.		
2.		
3.		
4.		
5.		
6.		
7.		
8.		

Name ______________________________

Exercise 19: Cost-Benefit Analysis

1. Use the Excel cost-benefit analysis template (template.xls) on your project disk to create a cost-benefit analysis worksheet for your proposed system.
2. In this template, several placeholders are provided for development costs, production costs, and benefits. Type over the placeholder (e.g., "Dcost1) in the Input section of the worksheet, entering your own data (e.g., "5 Pentiums with 16 MB RAM"). Do the same for all the zero values in the Input section. **DO NOT ENTER ANY DATA IN THE CALCULATIONS/ OUTPUTS SECTION!** All data in the Calculations/Output section are computed from your Input and Assumptions data.
3. Use the Assumptions section to perform a "What If ...?" analysis, entering different values for the following assumptions:
   - Sensitivity factor (Costs): A sensitivity factor of 1.0 indicates that you are certain of your cost estimates. A sensitivity factor of 1.1 indicates how the cost-benefit analysis will be affected if costs are 10 percent *higher* than your estimates. A sensitivity factor of 0.9 indicates how the cost-benefit analysis will be affected if costs are 10 percent *lower* than your estimates. A sensitivity factor of 1.2/0.8 yields the effects of actual costs being 20 percent higher/lower than your estimates.
   - Sensitivity factor (Benefits): The same sensitivity analysis can be performed on your benefits estimates.
   - Annual change in Production Costs: Enter a value here to indicate the annual change (e.g., 10 percent) in production costs. Enter your value as a decimal (e.g., .10 for 10 percent increase; -.05 for 5 percent decrease).
   - Annual change in Benefits: Enter a value here to indicate the annual change (e.g., 10 percent) in benefits.
4. Run several scenarios, printing the resulting worksheet for each scenario.
5. Attach your printed worksheets to this form.
6. In the space provided below (or on another sheet of paper attached to this form), comment on what your analysis reveals about the sensitivity of costs and benefits, i.e., do your estimates have to be very accurate to ensure that the project is economically feasible?

Team Member Names ____

Exercise 20: Construction Plan and Installation Strategy

1. Prepare a Gantt chart of the activities required to construct the system you designed.

Tasks	Resources								

2. Choose and briefly describe an installation strategy for converting to this system. Justify your choice.

Name ______________________________

Exercise 21: Test Specifications (page 1 of 3)

Complete a test specification for each of the interfaces -- report, data entry screen, and menu screen -- you designed for Exercise 16. Include at least 5 test conditions for each interface, and provide test data (where appropriate) against which to test your interface.

Test Data:

Interface: ______________________________

Condition #	Condition Name	Test Steps	Expected Results

Name ______________________________

Exercise 21: Test Specifications (page 2 of 3)

Complete a test specification for each of the interfaces -- report, data entry screen, and menu screen -- you designed for Exercise 16. Include at least 5 test conditions for each interface, and provide test data (where appropriate) against which to test your interface.

Test Data:

Interface: ______________________________

Condition #	Condition Name	Test Steps	Expected Results

Name ______________________________

Exercise 21: Test Specifications (page 3 of 3)

Complete a test specification for each of the interfaces -- report, data entry screen, and menu screen -- you designed for Exercise 16. Include at least 5 test conditions for each interface, and provide test data (where appropriate) against which to test your interface.

Test Data:

Interface: ______________________________

Condition #	Condition Name	Test Steps	Expected Results